Patrick Moore's Practical Astronomy Series

Other Titles in this Series

(Continued after Index)

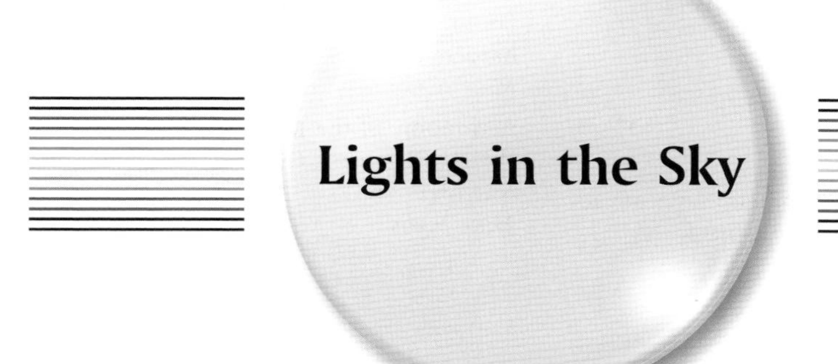

Identifying and Understanding Astronomical and Meteorological Phenomena

Michael Maunder

Michael Maunder

British Library Cataloguing in Publication Data
A catalogue record for this book is available from the British Library

Cover illustration: The Sword of Orion with Takahashi refractor. Courtesy of Jon Christensen.

Library of Congress Control Number: 2006940070

Patrick Moore's Practical Astronomy Series ISSN 1617-7185
ISBN-13: 978-1-84628-562-2
e-ISBN-13: 978-1-84628-761-9

Printed on acid-free paper.

9 8 7 6 5 4 3 2 1

Springer Science+Business Media
springer.com

Contents

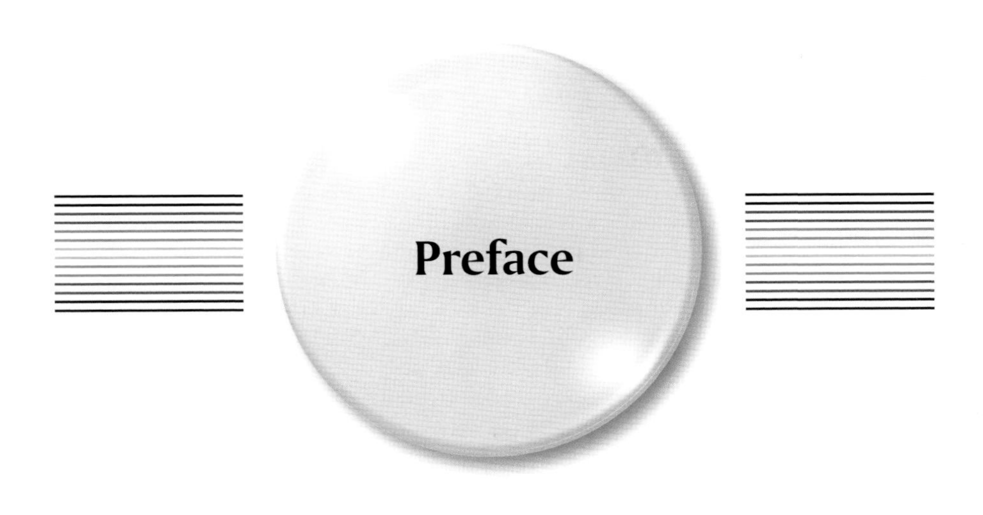

Preface

How well I remember days BC. That is, before Cars

The skies were always clear and the weather fine.

The way I recall it as a kid is as sunny in summer and as clear crisp and clear cold in winter for tobogganing, with the stars afterward to light you on your walk home at night, in complete safety. No cares at all, no cars running you down.

Why do I mention this? Quite simply because it highlights (pun intended) everything that's wrong today in our study of lights and glows in the sky. It boils down to our obsession with the car.

Children today, that's the majority living in or nearby conurbations, have no comprehension what it is like to live in unpolluted skies and appreciate the sheer delight(s) of looking up into a clear day or night sky and (actually) looking, as distinct from gawping, at the miraculous wonders in view. If you didn't know what that strange glow or apparition was in the sky you went to an elder and simply asked. That's how you learnt in those days from your elders who were, for all intents and purposes, part of your extended family.

Such knowledge I acquired inspired me to inquire further and eventually led to a science career and to writing about all these marvellous things in view that others seem to miss or ignore.

Too many children will never learn about glows in the way I did, by simply looking up. Even as late as my teens, half a century ago, it was quite normal at my school in central Croydon in the United Kingdom to look up and see the Milky Way and up to nine Pleiades. Today, the place is a huge shopping precinct, typical of most towns worldwide. If you can ever get a view of the Moon at all it's a weird colour except on the rare night.

However, that was before the Clean Air Act, when it often happened during a good Peasouper fog (Foggy London Town, no less) day that the sun vanished as well and us short-in-stature lads often had difficulty seeing the curb at our feet in daylight, the smog was so bad.

My point is that even that gross level of atmospheric pollution only occurred when the conditions were right; today the smog is a lot thinner but just as pernicious to health and our enjoyment of things to see above.

How did I get interested in glows? That's something very difficult to reproduce in today's overprotected culture but here it is. My first recalled memory on this topic was the glow of searchlights in the sky, criss-crossing the Milky Way, a stark contrast in intensity but both visible in total blackout; if you were lucky a glimpse of a barrage balloon blinking in and out, or an aircraft which then got caught in several beams. One never forgets a plane coming down in flames, the ultimate horror as one thinks about it now.

I suppose it was this essential and easy need to keep looking up and avoid the first missile attacks that spurred my later interest, as the red glow out of the tail of a "Doodlebug" at night certainly grabbed one's attention and memory.

By looking up at contrails by daytime, a gradual interest in astronomy matured and was kept kindled for the next few years by Dr. J. G. Porter's regular monthly radio broadcasts. (TV was a 4-in. green glow gizmo that someone had down the road.) Then in the 1950s one of his broadcasts did the trick. His enthusiasm simply meant you had to go out to see what he described. That's when I saw the amazing sight of nearly all the planets strung out in a line, in a gorgeous sunset sky. I was hooked.

This book is an attempt to do something positive and pass on a little of what I've experienced before I become a glow in the sky myself.

Michael Maunder
Alderney, August 2006

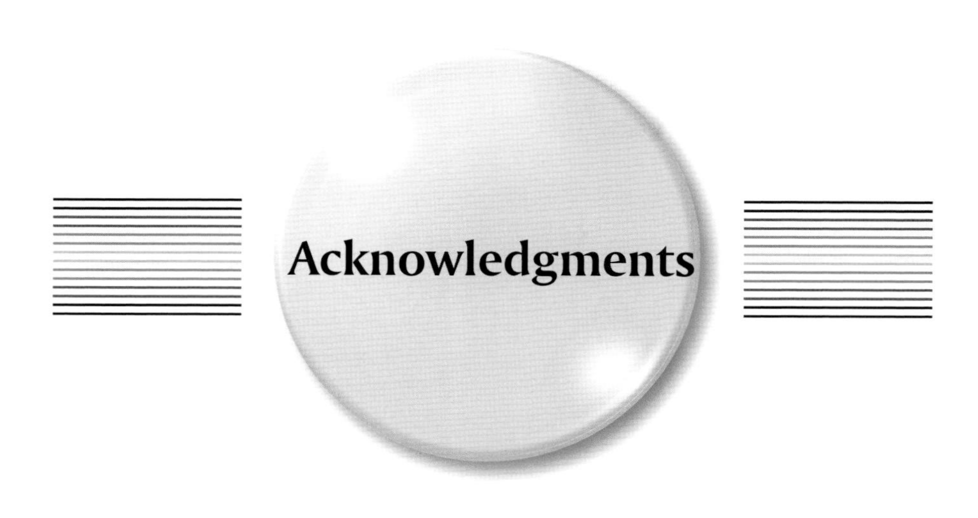

Although this book doesn't break any records for a long gestation period, it certainly was for a long time coming. If it hadn't been for the enthusiasm and encouragement in the early days of Patrick, now Sir Patrick Moore, I'd never have started at all.

His vast knowledge and expertise daunted me from the starting. It was the sheer complexity of the topic that I've commented on in the Introduction that delayed my starting, as each subtopic he suggested alone genuinely required a book alone. In the end I did get underway and this book is dedicated to him as a written token of my appreciation.

Another person responsible for applying the right sort of pressure is John Watson. Without his considerable publishing skills and devotion to setting me on the right track, the text would have fallen by the wayside long ago. One topic I couldn't handle from lack of expertise is the loaded question of Aliens and UFOs. John kindly offered to deal with that and I've incorporated it as a fitting rounding up of the subject.

My grateful thanks to Sir Patrick and John for their fortitude in the face of such a slow starter as myself, and to my wife who is still waiting for all those DIY jobs about the house.

Michael Maunder
August 2006

CHAPTER ONE

Introduction

One of the curious paradoxes of the car has been the way it's freed us to live where and whenever we like, whilst at the same time imprisoning us.

Instead of children walking, the car is brought out and the poor things lose out on every count. Health factors aside, there's a fundamental lack of curiosity in the environment about them because they can't be allowed to stop and inquire on the way of anyone like myself with a lifetime's experience to pass on.

Those cars making unnecessary journeys add yet more pollution and that reduces or totally eliminates many of the delights to be described here. Half a century ago, cars did indeed pour out very nasty exhausts, but only a fraction of coal in domestic stoves. The modern car is said to be "clean." The problem lies in the fact that there are simply too many of them and in sum total the atmospheric pollution hits you with a double whammy.

First, the exhaust pollutes the air and leads to modern variants of the famous Los Angeles Smog, which markedly cuts down air clarity. Second, at night, motorists insisted that roads had to be brightly lit. That streetlight pollution quickly killed off any hope of seeing even the brightest of stars. Their daytime and night exhausts created a fine haze which throws light back into your eyes by the obvious phenomenon of "backscatter." As we shall see, simple light pollution isn't the whole answer to the loss of night skies, the car and household exhaust have a lot to answer in destroying sky clarity from the wonders still on view.

However, it's not all doom and gloom, and pollution can give some very pretty glow phenomena in their own right as will be seen.

The topic of lights in the sky is so vast and intricate that no attempt has been made to be comprehensive. The intention here is to select some of the more common, or interesting, aspects and to comment on them in two ways.

We start most chapters with a brief introduction of what's going on in "Mechanism", and then follow that with some ideas of what to do in "What to Observe".

Suggestions of what to see or at least strive for are much more important in a book within a series of practical topics than the minutiae of detailed theory and complexity. For that reason, only some brief references have been made to other books in this series. The reader will obtain much more data from the web if that's what they need to follow up the theory. There is a very practical reason for that leap of logic—all of these individual topics and subtopics are so complex in themselves that a book solely devoted to that limited field is entirely justified. Indeed, many such books have been written on just those items.

Many lights in the sky can be seen throughout the year and at any time of day, with some at night. Many of these phenomena are so rare that I've not seen them, and others so "trivially" common that nobody gives them a second glance. A pity as we'll see.

Wherever possible I've used my own illustrations as that further tells the tale of my journey through life. Perhaps the most interesting fact I learned in doing that was very salutary. I'd always thought of pictures as just "pretty pictures" not their having some science hidden within them. Who'd have thought a 1970 picture of Niagara Falls contained the secret of a rainbow?

Where other pictures have been used, the author is credited.

Because some of the lights are so faint, they hardly merit that name. Throughout this book I'll describe them as "glows"; a fitting description when you "glow" with pride at spotting your first rare one.

These following Illustrations 1.1 to 1.4 help to understand many of the fundamental principles involved and how the optical and atmospheric phenomena fit together. Reference back to them is made where necessary in the later Chapters.

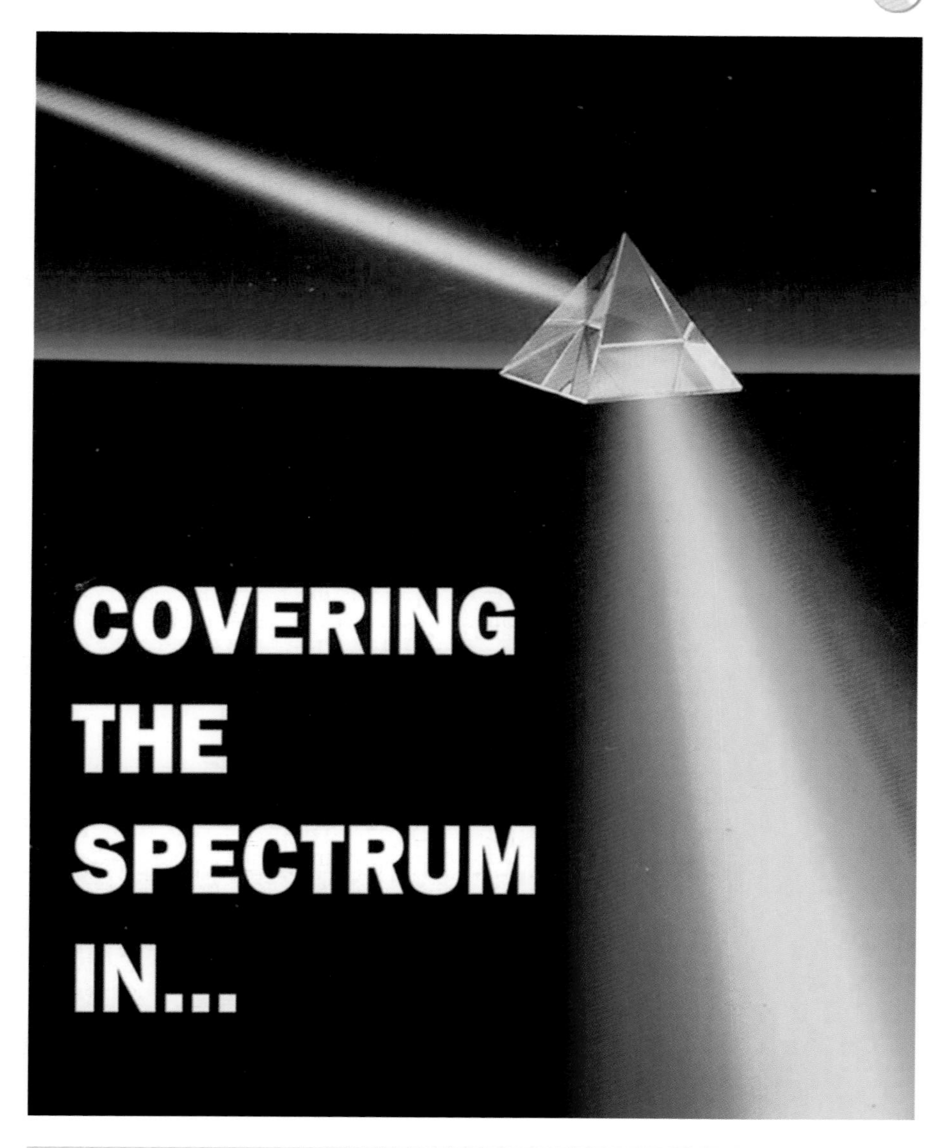

Figure 1.1. ... **EMBARRASSMENT!** By far, the most common mistake is illustrated here. As long as you remember that blue/violet light is deviated more than red, most of the lights in the sky begin to make more sense.

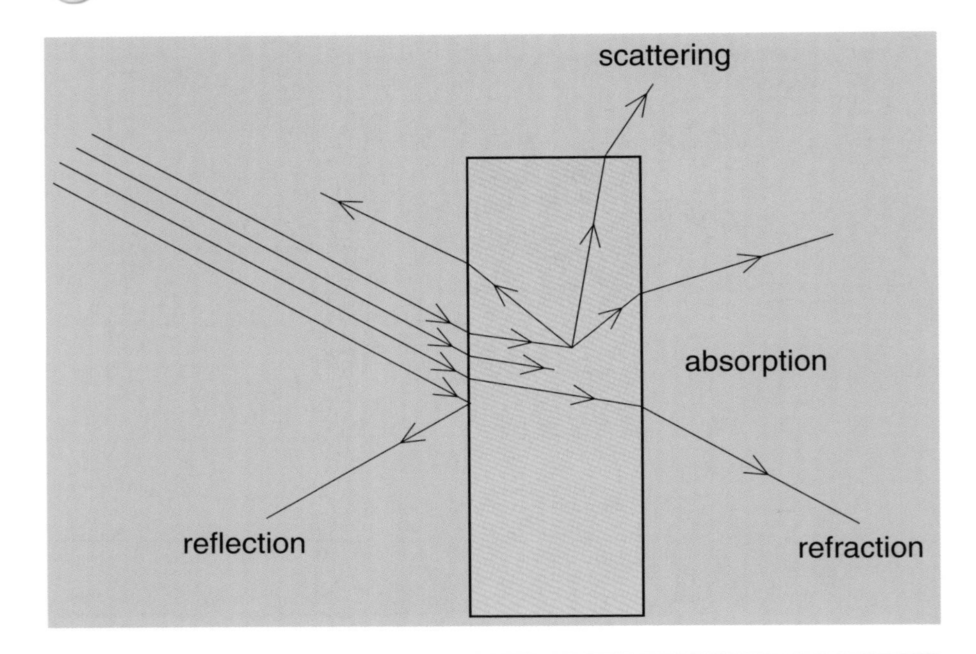

Figure 1.2. Light paths. The various ways light can change direction. It can be absorbed completely, reflected back at an equal angle of reflection, refracted and carry on in the same direction although deviated slightly, or scattered in any conceivable direction. When light is within the object, each of these light paths can generate any of the other modes.

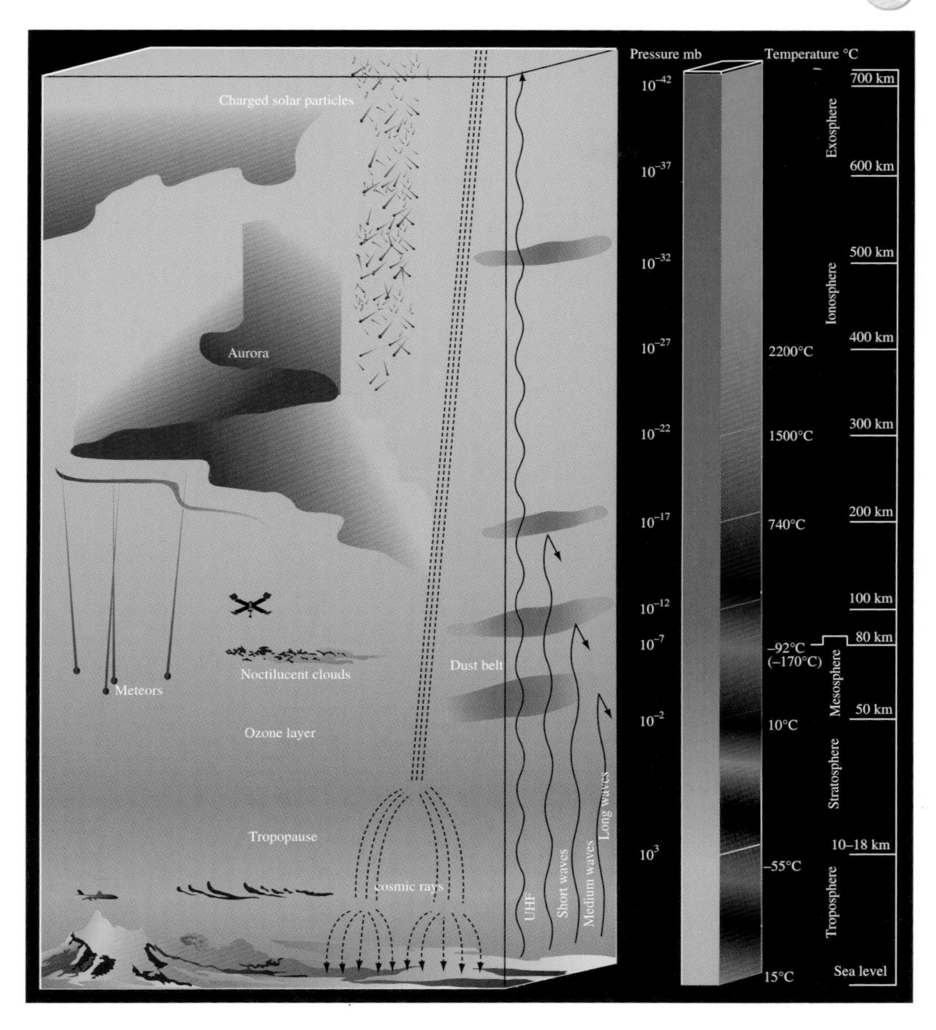

Figure 1.3a. *Atmospheric layers* We live on the ground, but above us the air settles into distinct layers, each with its own composition and properties. The boundaries constantly change height with seasons and time of day, and that with true space is never static due to solar wind and other radiation creating its own atmosphere around the sun.

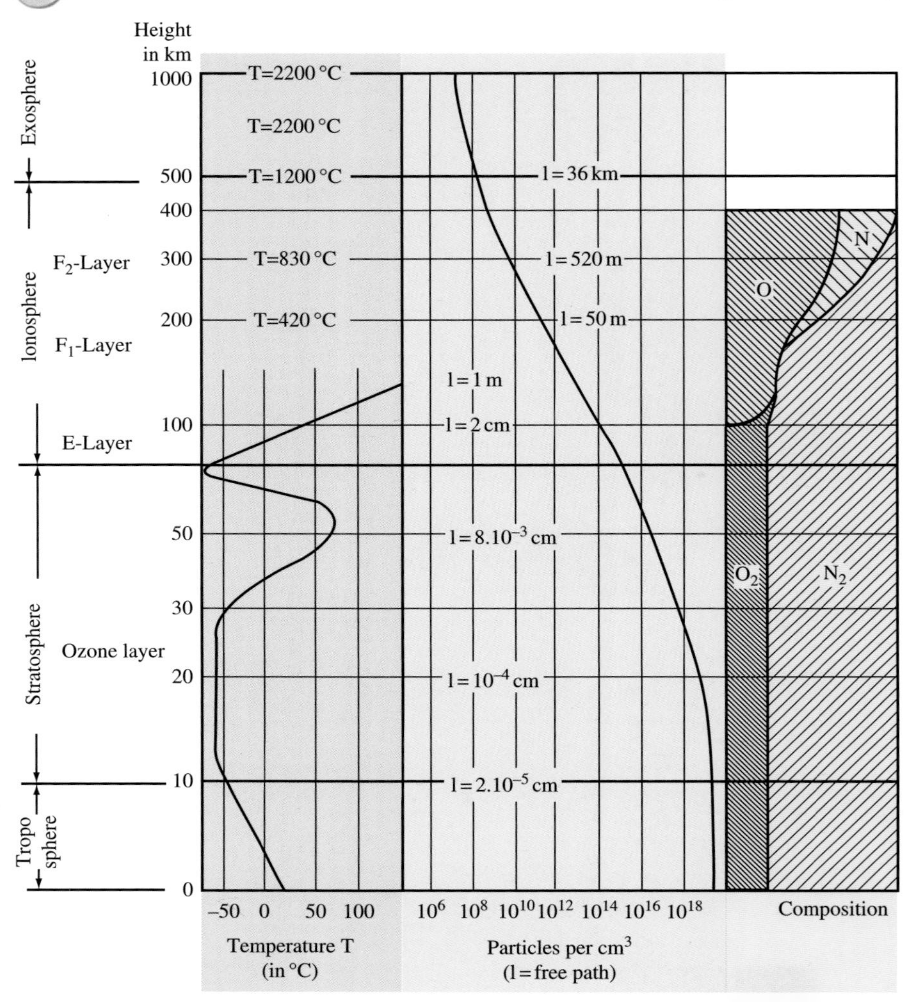
Height
in km
1000
500
400
300
200
100
50
30
20
10
0
Exosphere
Ionosphere
Stratosphere
Tropo sphere
F2-Layer
F1-Layer
E-Layer
Ozone layer
T=2200 °C
T=2200 °C
T=1200 °C
T=830 °C
T=420 °C
l = 36 km
l = 520 m
l = 50 m
l = 1 m
l = 2 cm
l = 8.10-3 cm
l = 10-4 cm
l = 2.10-5 cm
N
O
O2
N2
–50 0 50 100
Temperature T
(in °C)
106 108 1010 1012 1014 1016 1018
Particles per cm3
(l = free path)
Composition

Figure 1.3b.

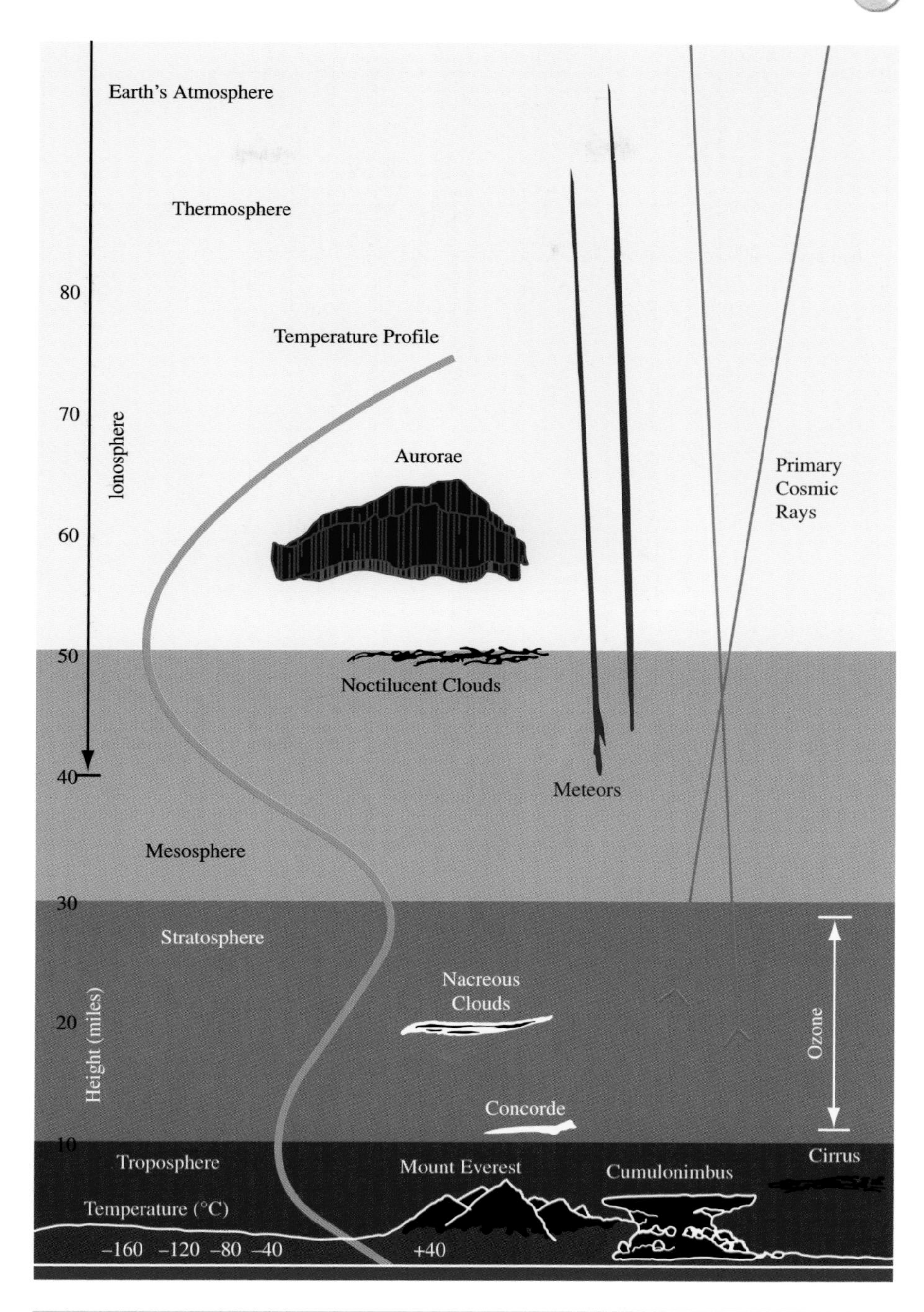

Figure 1.3c. *(Continued)*

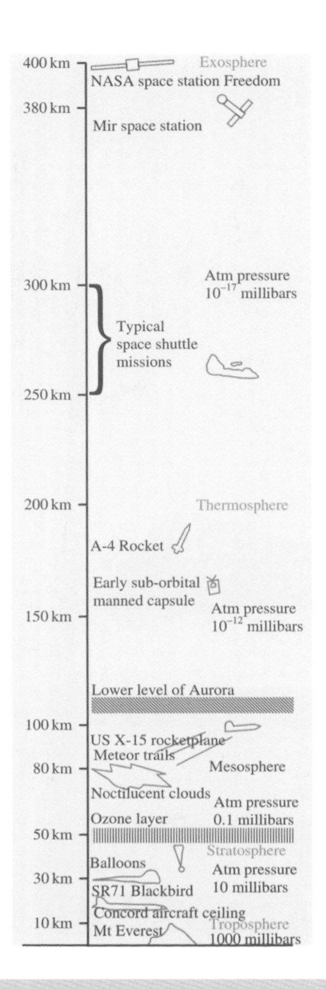

Figure 1.3d. *(Continued)*

Figure 1.4. Spectrum at shuttle launch site.

CHAPTER TWO

Light Pollution

All too often, light pollution, much better called sky glare or skyglow, is a totally underrated problem. And for a quite simple reason it's underrated, as most people live in a brightly lit urban environment where they think this light level is normal because that is all they ever experience. Of all the glows considered in this book the glare from urban and road lighting has become the only one to be guaranteed of a good showing each and every night.

The problem doesn't stop there in isolated country situations. The recent obsession with safety has led to a major industry in the so-called "security lighting." It is rather sad that this lulls the owner into a false sense of security. Most of the safety lights installed are much too bright and become counterproductive by blinding anyone outside the glare who might otherwise see an intruder up to no good.

Fortunately the public perception of the general problem of light pollution is swinging into a better realization of the real situation. The impetus for change started in several countries at the same time, but by far the biggest of these was the campaign run by "Sky & Telescope" in the early 1990s, and from the United Kingdom by the British Astronomical Association.

Astronomers are not asking for a total light switch-off but correctly designed installations. This not only saves energy, but also directs a lot more (effective) light, where it does the real job.

Governments have already passed light pollution Laws, and others are also considering Legislation to curb the nuisance aspects of light pollution, and quite a number have already done so in the way roads and motorways are lit, the major problem outside conurbations. Good examples are to be found in Southern England, where driving down the motorways had become a lot safer. Lighting is now being directed downward onto the road and not as all too common in the past—in all directions, straight into the driver's eyes to cause dazzle and then very rapid eye fatigue.

This is the nub of the problem; how the light is used or directed, not the amount per se.

2.1 Light Trespass

An almost totally new concept in recent European Legislation, but always a familiar term in America, "trespass" has come to pass. It's a lot easier to define and recognize.

The legal definition runs along these lines: "Nuisance light that crosses property lines to fall where someone objects." Obvious cases are much easier to prosecute when a new installation is under review and include bright light(s) into a bedroom window. The potential medical condition of sleep deprivation is not a new one and that makes objection so much easier.

Arguing against light onto your observing site is a different and more difficult one. Test cases will occur but at the time of writing the way forward is not obvious but a compromise for agreed "On & Off" timings might work. Floodlights alight all night at a baseball or football stadium when they're only needed when games are in progress seem logical "no go." Golf Driving Ranges lit all night are another pet hate for most astronomers.

The chances are there's something similar near you.

2.2 Waste Lighting

All too common is this, the related problem to trespass. I define it as light that pours needlessly upward or horizontally out of poorly designed fixtures.

The light engineers and manufacturers are already adopting Codes of Practice, where the rule of thumb goes that only rays directed 20 or more degrees below horizontal generally serve any useful purpose.

Globe lights at supermarkets are the worst light wasters, but what's your experience?

2.3 Glare

Again, we can think of this as anything arising from either or both these main sources. It's light going directly into your eye from a bulb brilliant enough to interfere with your vision.

The light doesn't have to be unusually bright to create this effect. I'm lucky and chosen to live on a small island where the streetlights are few and reasonably well designed, and I've always preferred to walk without torchlight to preserve night vision. Still keeping in excellent condition well into my dotage.

Even at a mile away (well over a kilometer), the car park or security light from my home apartment block creates dazzle and I cannot spot any road surface unevenness, a distinct hazard. Walking down the final 100 yards or meters of road without walkways/pavements is a completely frightening business since *none* of the road surface is seen. And that's from a single modern low-power bulb of just a few watts. In a normal urban road that bulb would hardly be noticed, so the glare problem boils down to one of the ambient environment itself. The better the environment, the more obtrusive the glare becomes.

Glare creates an ugly, uncomfortable, sometimes a dangerous environment. The bulb is being shielded, and as the compromise arrived at is that it goes off at midnight, I cannot object to that bit of common sense. The compromise permits real sky studies from nearby when nobody else is about, for the rest of the dark hours.

2.4 Skyglow

Skyglow is the culmination of all these factors and is increasingly not just a totally urban problem. It is defined as artificial light scattered back down toward the ground from the atmosphere.

This problem is linked with atmospheric pollution because the worse the muck in the air the worse it becomes, and is considered in more depth in Chapter 3. The only entirely natural skyglow arises from light reflected back from cloud cover. We're all familiar with that.

2.5 Security Lights

Contrary to the myth or to most people's unconscious belief, directly viewing a light bulb does not increase your safety or security. Most installations are a total waste of time and money.

Any burglar you want to deter is not going into your premises straight at the bulb height (like a moth). He's not attracted that way to your glow but down at ground level, which is where the only illumination should be.

The basic rule of good outdoor security lighting is that no one should ever see a bright bulb. It's so much more effective to go for a passive infrared system, which is then used to switch on a temporary searchlight; even better still, why not use an infrared sensor for video and catch the burglar on video tape, as it were, literally, red-handed? Law Enforcement Officers can take over with that tangible evidence.

Light pollution is an ever-evolving subject, and the way ahead does seem to be much "clearer" than only a short time ago. Non-astronomers are beginning to understand the importance and need for a better environment. An appreciation of astronomy does seem to lend itself as a good tool to understanding these other aspects of the pollution going on in the wider environment.

Figure 2.1. *Arizona, 1994* In perfectly clear skies, light from the full moon doesn't stop photography of the natural lights in the sky-stars.

Figure 2.2. *Helen Gorge, Australia* In the Southern skies, another full moonlight star picture.

Figure 2.3. *Hyakutaki at La Palma* Although designated a light-free zone for astronomers, La Palma's town lights show up in this picture.

Figure 2.4. *LA, 1970* Even in "LA Smog" days, the skies could still be seen in the modest street lighting of the 1970s.

One saving grace from the hype about Global Warming and unsettled political crises is the upward spiral in energy costs. That helps us in two ways. First, by encouraging more efficient lighting systems, or even to switching off altogether; and, second, by creating situations where the energy supply is inadequate, leading to power cuts.

Perhaps it's wishful thinking, but here's to the next power cut with everyone out looking at their new and totally dark sky. Even with some natural light 'pollution' from the Moon you'll still get a beautiful experience of what you've been missing, as Figures 2.1 and 2.2 demonstrate, and Figure 2.3 shows pollution where it isn't supposed to be, and 2.4 where it can be absent in good nights.

"May the Force be with you."

CHAPTER THREE

Air Pollution

Writing on this topic via a book devoted to glows in the skies presents a few problems.

3.1 Mechanism

Anyone even casually interested in environmental issues will already know more than enough to pass onto the next section, whilst those disinterested or dismissive are inclined, or even more likely, to do the same.

My guess is that few will be unaware of the huge number of problems facing us. Figure 3.1 is an example, as is Figure 3.4.

The health hazards we face are hammered home more and more by governments and the direct link lies in Los Angeles and London leading the way to a variety of Clean Air Acts. First, some different forms of atmospheric pollution and how they affect our lights in the sky are discussed.

3.2 Dust

Section 7.4 of Chapter 7 goes into the effects this can have on our appreciation of the natural lights in the sky. These basic factors must be considered first.

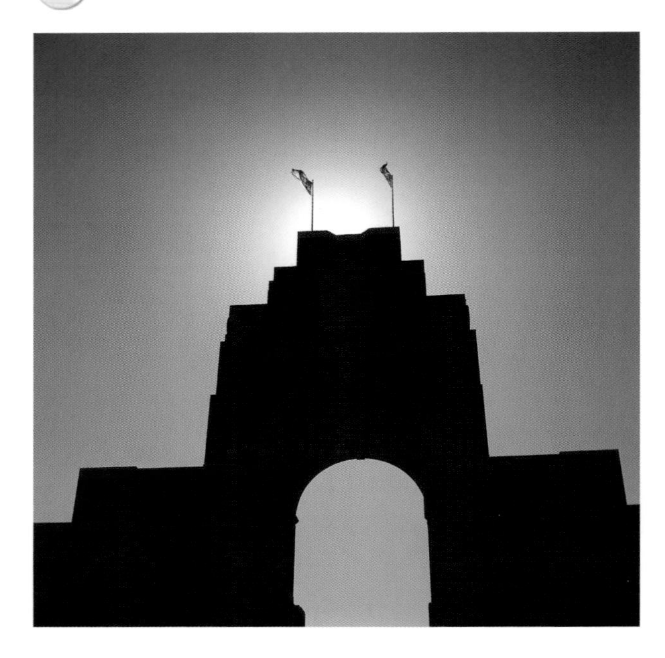

Figure 3.1. *WWI Memorial, 2003* Industrial haze in Europe enhances the dramatic symbolism of the WWI memorial. A rare example where atmospheric pollution has benefited a picture.

3.3 Desertification

Dust in the atmosphere is an entirely natural feature that's gone on from time's beginning. This stuff is derived from anywhere the ground is dry enough to allow winds to pick up particles and carry them away (Figure 3.2).

It seems that every day some newspaper reports how the world is going into crisis with water shortage and the extending deserts. Africa has seen the Sahara extend since Biblical times, about the same period that many of the South American Civilizations ran into difficulties.

The deserts encroach ever further from their traditional range. Figure 3.3 illustrates this process in action as it affects America from a different continent.

3.4 Particulates

The problem as far as it affects our interest in lights and glows in the sky is intricately and profoundly bound up with a new range of atmospheric pollutants. The type of pollutant changed, it has not gone away at all with coal smoke. Nor has it with really stringent car-testing and exhaust filters.

These easily visible air pollutants in the earlier part of the twentieth century have been replaced by other much more insidious "air-additives", amongst which are particulates from diesels and similar exhausts. These particulates are much too small to be seen with the unaided eye until they settle out as a general smear on anything around.

Figure 3.2. *Dust devil Kenya, 1976* A very minor example of how dust gets into the atmosphere, then into its higher reaches.

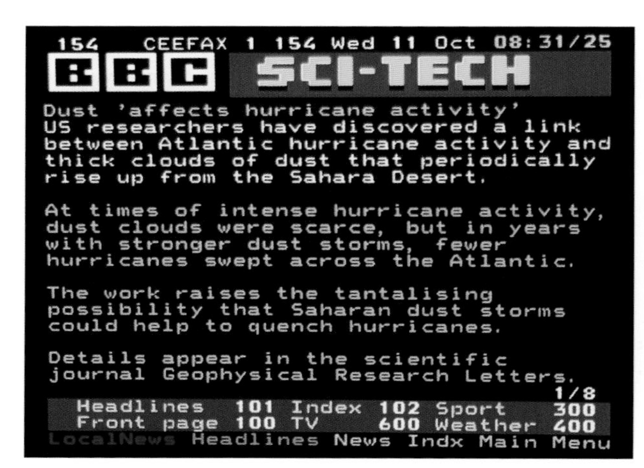

154 CEEFAX 1 154 Wed 11 Oct 08:31/25

BBC SCI-TECH

Dust 'affects hurricane activity'
US researchers have discovered a link between Atlantic hurricane activity and thick clouds of dust that periodically rise up from the Sahara Desert.

At times of intense hurricane activity, dust clouds were scarce, but in years with stronger dust storms, fewer hurricanes swept across the Atlantic.

The work raises the tantalising possibility that Saharan dust storms could help to quench hurricanes.

Details appear in the scientific journal Geophysical Research Letters.

1/8

Headlines 101 Index 102 Sport 300
Front page 100 TV 600 Weather 400
LocalNews Headlines News Indx Main Menu

Figure 3.3. *Sahara dust, October 2006* This typical report of a huge storm illustrates a major factor in weather and climate change on the other side of the world. Iron in the dust is a major food source for pelagic (oceanic) plankton. © BBC

Figure 3.4. *West Coast California, 1975* Scenes like this led to antipollution laws, but worldwide the problem persists with new emissions, as in Figure 3.5.

When diesels first came in, their dirty exhaust was only too obvious. What's happened recently is that the exhaust has been cleaned up visually, but all the minute carbon and part-burned fuel that is left cannot be entirely removed. Whilst it might appear entirely transparent to the eye, there remain these unbelievably small particulates. Their average size is $< 10\,\mu\text{m}$, hence their name PM_{10}.

The main problem with these particulates is their very size. They're of a size to cause lung problems, but much more important from our point of view, although invisible to the unaided eye they can affect sunlight and starlight by scattering it, or diffracting it, or both. This is a feature of particles this small, described as colloidal. When mixed with a cocktail of noxious gases like ozone and nitrogen oxides, the scene is set for Los Angeles smog (Figure 3.4). The problem is still with us today (Figure 3.5).

Figure 3.5. *UK Pollution, August 2004* Three decades later, we still see severe atmospheric pollution over the United Kingdom in a hot summer.

That's the core problem in our concerns with car exhaust. It isn't the major contributor to carbon dioxide emissions. It's this much more pernicious particulate matter which should concern us particularly because of the accompanying host of organic chemicals and nitrogen oxides, all of which contribute to the overall atmospheric pollution in urban environments.

Many more cars, although described as low emitters, in toto send more than enough exhaust into the atmosphere to completely wipe out any chance of seeing the Milky Way and all but the brightest stars. That can happen on any day in huge tracts of country when the air is settled and the wind doesn't disperse the stuff. When wind does spread the cocktail into the upper atmosphere some interesting effects arise.

As we'll see in later Chapters, these "invisible" pollutants play a crucial role in some of the rarer glows in the upper atmosphere, and the weather in particular and the climate in general.

3.5 Smoke Dust

The other marker to dust pollution has to be how we burn fossil fuels. That's well covered in the literature and is under some form of control.

Smoke dust here is not from industries but from forest fires. Some are natural, some entirely man-started, be it accidental or deliberate. Natural forest fires have been on the increase in recent years, in parallel with forest clearance. The particulates generated have had a devastating effect on the weather over the whole of continents and on the health of the population. Figure 3.6 reports a natural event, whereas the caption to Figure 3.7 shows how some Countries have a long way to go to exercise some form of control of atmospheric dust and their associated "invisible" pollutants.

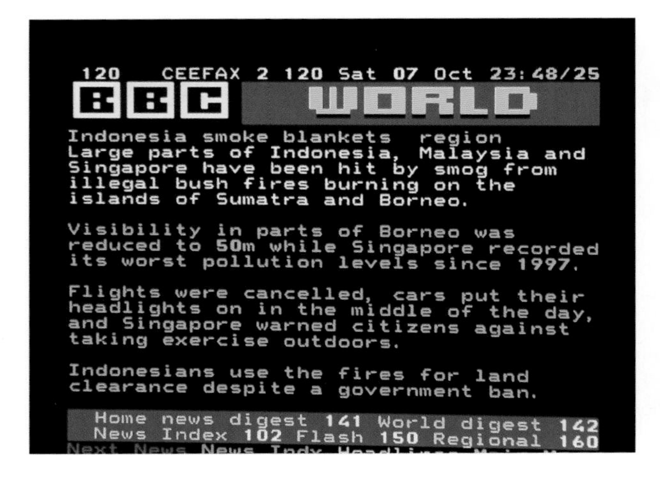

120 CEEFAX 2 120 Sat 07 Oct 23:48/25

BBC WORLD

Indonesia smoke blankets region
Large parts of Indonesia, Malaysia and Singapore have been hit by smog from illegal bush fires burning on the islands of Sumatra and Borneo.

Visibility in parts of Borneo was reduced to 50m while Singapore recorded its worst pollution levels since 1997.

Flights were cancelled, cars put their headlights on in the middle of the day, and Singapore warned citizens against taking exercise outdoors.

Indonesians use the fires for land clearance despite a government ban.

Home news digest 141 World digest 142
News Index 102 Flash 150 Regional 160

Figure 3.6. *Southeast Asia smoke, October 2006* Deliberate smoke pollution has a drastic effect on nearby skies and climate. © BBC

Figure 3.7. *Yangtze colliers, 1997* Scenes like this will continue as China's the largest coal burner, and releaser of firedamp (methane) 23 times more potent as a greenhouse gas than the carbon dioxide formed in direct burning. Radon released in coal mining has an unaccounted effect on cloud formation.

3.6 Contrails

Another man-made effect on the weather over quite wide areas on the planet is aircraft contrails. Areas well away from popular traffic routes still enjoy crystal clear skies, whereas huge areas of America and Europe sometimes have nothing but these clouds' cover from horizon to horizon. Whatever the long-term effect on the *climate* turns out to be, the effect on the *weather* is patently obvious (Figure 3.8).

Some recent research indicates contrails can be reduced by almost a half by the simple expedient of flying aircraft as little as 6,000 feet lower. Quite what effect that has on fuel consumption is another story, yet to be unraveled.

We get some extremely pretty sky glows from these newly created clouds, quite unknown a century ago!

Figure 3.8. *Persistent contrails* An all too familiar scene, possible portent of bad weather to follow, blanking out all lights in the sky, except lightning (Chapter 13).

3.7 Volcanic Dust

Volcanic activity is a prime factor for dust in the atmosphere. Sometime soon, maybe before this book reaches you, a big one will turn out to be a main cause of the climate turning cold, very cold. The prospects for not seeing any sunlight at all for years don't bear thinking about if Yellowstone blows. (But see also Figures 3.9–3.11).

3.8 What to Observe

Without the inspiration of our natural lights, environmental awareness is blighted and the true impact not considered at all.

If you never see the real or natural night sky, how can you appreciate it? It is almost like describing the blue sky to a blind-from-birth person or a violin's sound

Figure 3.9. *Mt. Erebus in Antarctica, 1998* This natural air polluter contributes to global climate balance. Rarely seen, therefore this active volcano is "out of mind."

Figure 3.10. *Mt. Etna in Sicily* Regularly erupting, this volcano has little effect outside the immediate region. The effects on clouds are more remarkable.

to a deaf one. Their brains are programmed to make sense of what they experience but never the whole.

This applies to anyone living in a polluted environment from birth, never experiencing a clean one; the interpretation of what's important differs from those on the outside. Both take their situations as normal, and if those in cities are in the majority, the preference for pollution, be it car exhaust and particulates or light, can rule.

The more subtle effects of atmospheric pollution are becoming much more familiar as they are more widely publicized. I sincerely hope that by the time these views appear on the bookshelf, a lot more progress will be made.

If not, some of the more bizarre glows in the sky due solely to air pollution will become much more familiar.

Figure 3.11. *Krakatoa erupts by me, 1988* I'm lucky to be alive as the eruption an hour or so later went up to stratosphere. The sunset skies were impressive as we sailed away! Visitors are no longer allowed without permits.

CHAPTER FOUR

The Weather

We can't jump straight into this topic without making the standard joke: "Everyone talks about 'Climate' but what you get is 'weather'."

4.1 Mechanism

4.1.1 Cloud Types

Throughout the book and later in this Chapter we'll need to refer to the clouds and how they can become part of the display. For the time being, the diagram shown in Figure 4.1 is a simplified summary of those of immediate interest.

Where a specific cloud type is involved, the features of relevance at that time are given.

4.1.2 Warm Front

One stage on from the cloud types is a basic knowledge of the weather patterns. The one of most relevance here is the warm front and the colder and cleaner air that follows behind. Obviously, real situations are much more complex than that. Study Figure 4.2 and refer back as it crops up later.

Figure 4.1. *Cloud types* For clarity, only the main varieties of interest to us are shown.

4.1.3 Climate Change

At the beginning of the twenty-first century the jury is out, still deliberating whether or not the current trend is toward a warmer climate worldwide. The evidence seems overwhelmingly that it isn't a local affair. Where many debates still take place is on the more knotty evidence that the sole contributory factor is manmade through greenhouse gases, specifically in the carbon dioxide increase from fossil fuel burning.

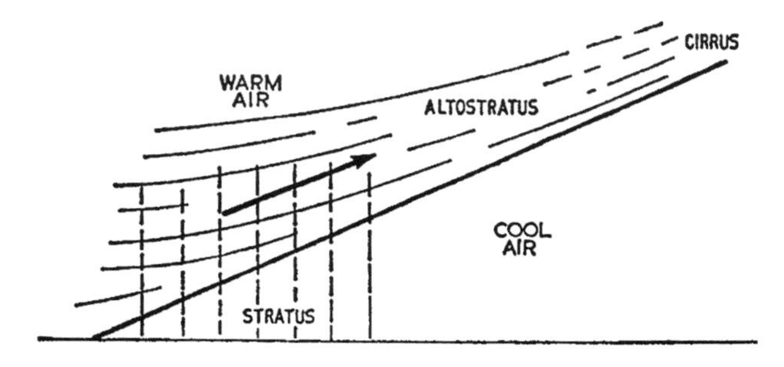

Figure 4.2. *Warm Front* A red sky in the morning needs this warm front arriving during the night. It is also obvious why the sunsets are a quite different phenomena.

I have my doubts that this is the real villain of the piece. It is certainly ***a*** contributory factor but the more I researched for this book, the more I've become convinced that many of the glows we see in the sky attributed to global warming arise from entirely natural features.

4.1.4 Speed of Change

The evidence that I've collated also clearly indicates that earth's normal history is one of the dramatic switches in climate regime. Not in millennia, nor centuries but quite often in a matter of days. We've been incredibly lucky to be living in a period when the climate has been unusually stable. The main conclusion is that we'll simply have to get used to whatever nature has in store for us because current thinking on how to remedy the situation hasn't much of a clue on how to cope with the true forces at work.

A major volcanic eruption, for instance, will certainly concentrate the mind in the other direction as we experience a catastrophic rapid cooling. Krakatoa was a baby compared with Biblical accounts of Santorini, for instance.

Topping the list has to be two unpredictable factors, which are variability in the Sun's energy output and the decline in the earth's magnetic field. Both of these cause the upper atmospheric glows, which have been on the increase. Close monitoring of these glows will greatly assist resolving that side of the debate.

4.1.5 Solar Variation

Studying the Sun's activity is now a major R&D effort in most industrialized countries. They accept that the radiation from the Sun is extremely variable, and sudden flares and storms on its surface profoundly affect and interfere with the earth's magnetic field and with it all electrical equipment.

Some very important satellites have been launched recently with the sole purpose of anticipating solar effects. In turn those studies will help elucidate quite how far the present warm spell is influenced by the solar variations.

Visible light is very constant, but the ultraviolet and particulate matter fluctuate by many hundred- or thousand-fold. Some discussion is in the relevant Chapters on aurora and other upper atmosphere clouds.

4.1.6 Magnetic Field

At the time of writing this, in mid-2006, magnetic effects weren't fully taken into account in *global* weather models. The earth is rapidly reversing its magnetic field and that started a much sharper decline from 2 to 15 nT per year, coincidentally, about the time of the Industrial Revolution in the mid-1800s—where most "evidence" for man's influence is shown on carbon dioxide curves.

Tessler is a measure of the earth's magnetic strength, and the most drastic decline in its magnetic field has occurred close to the South magnetic pole, where debate on ice sheet melting is popular. What is certainly true is that the increase in cosmic ray and solar flux into that region can and does affect the upper atmosphere climate. Some satellites have already been damaged in those increased radiation fluxes.

With this more than 7-fold speeding up of the earth's declining magnetic field, the polarity is set to reverse completely well within the next millennium. The compass made today will show the pointer heading South (somewhere!). Long before that we'll be subjected to the full blast of solar and cosmic radiation. That's set to be a much more worrying aspect of the (wider) climate change than just the temperature; definitely something to watch on two counts:

1. Increasing lights in the sky in the upper atmosphere. True dark night skies worldwide could become a thing of the past.
2. Huge increases in hard and cosmic radiation reaching the ground. The ozone hole could become worldwide and permanent, letting in hard ultraviolet radiation A as well as B & C. Even Sunblock 200 will be ineffective.

4.1.7 Methane

Another, third factor in many, perhaps a majority, of the dramatic switchings of climate in geological history also indicates that the main upward driver for temperatures isn't carbon dioxide level. Charting on an annual basis, as distinct from centuries to aeons, carbon dioxide levels seem to follow, rather than precede, warming periods in an "interesting" percentage of such rapid climate change events. Indeed, in an equally surprising number of cases the true marker for warming is methane. Our quest for coal liberates more methane into the atmosphere than many of us care to admit, indicating that fossil fuel burning into carbon dioxide isn't the only true marker. Methane is approximately 23 times more efficient and can have an effect on some of the lights in the sky that will be discussed.

In that aspect, man is definitely affecting the climate. It's not what we burn that counts here but what *isn't* ignited, or radioactivity released as in Figure 3.7, plus whatever is liberated by the paddy fields, bogs, and animals.

4.2 The Atmosphere

Perhaps the most surprising feature of the atmosphere is the variability in it as one goes up into the true vacuum of space. We live on the surface and only experience weather within a very narrow band. Above that surface the atmosphere splits into at least four more distinct bands, each with its own type of weather and ability to exhibit glows. The vastly simplified diagram in Figure 1.3 will help explain what's going on.

Another most surprising feature is the counterintuitive temperature in some of the upper layers. Until we flew into the stratosphere, it was assumed that the temperature dropped steadily as one went up until the absolute cold of space.

Certainly this is true into the highest bit we scaled and climbed at mountaintops. Today we all know that at the top of the troposphere's weather band, making a boundary with the stratosphere, there is a sudden change in this steady drop, and the temperature remains reasonably steady around −55 °C. This is the appropriately named tropopause. Check this for yourself by information displayed in most modern jets during flight.

This boundary layer also contains the jet streams, which are a major contributory factor in the weather below. In the United States, these streams are always displayed on weather forecasts, and in other countries such as the United Kingdom it is rare, if ever. Knowing where the jet stream is at any time is of a huge help in predicting the type of weather, and with it the likelihood of glows.

The year 2006 turned out to be a classic example of this prediction aid. The jet stream over Europe was much further north than normal, resulting in much more settled conditions below latitudes around 52–53°. Weeks of exceptionally warm and still air led to a bumper crop of noctilucent clouds in most of mid-Europe. Scandinavia suffered lots of rainfall.

None of these atmospheric layers are ever static, and even their depth can vary over quite wide limits.

4.3 The Troposphere

The actual thickness varies according to where you are, that is thicker at around 6 miles (10 km) on the equator and shrinking to about two-thirds of this at the poles. How mountains, particularly at the South Pole, generate Katabatic winds and thereby influence weather patterns is a separate topic, not relevant here until we look at mirages and other upper atmospheric glows in some later chapters.

At ground level the air pressure is always changing and that becomes a good indicator in its own right of the weather at the time or to expect soon. The high air pressure is generally anticyclonic with clear skies, leading to balmy summer days and crisp frosty winters. These are the periods when we can get excited and expect the upper atmosphere and celestial glows to show up best.

Low and rapidly falling pressure invariably mean cyclonic conditions with cloud cover, our best expectation, and/or rain. Thunderstorms often occur at these times in late spring and summer when the weather patterns are changing most rapidly. The best that can be said for cyclonic conditions is that immediately afterward the best sky clarity is seen for marvelous views of the fainter glows in space. All the rubbish gets washed out of the air, and (in the Northern Hemisphere, anyway) cold clean air comes in from the unpolluted Arctic regions. In the Southern Hemisphere there are large ocean tracts with few inhabited regions close to the pole to complicate the issue.

I'm pleased to coin a new phrase or saying, "The north Wind doth blow. . .and we shall have glow!"

4.4 Stratosphere

Sitting immediately above the tropopause's dividing boundary, up to about 30 miles (50 km), is a critical region for our survival. It contains the ozone layer filtering out the bulk of harmful ultraviolet radiation from the Sun and space generally. How we see some of the more obscure glows is highly dependent on the state of the ozone layer. The thicker/denser the layer, the less ultraviolet gets through to influence events.

Contrary to common belief, the so-called "ozone hole" is not entirely dependent on human activity, and of late might be entirely unrelated after the Kyoto Protocol banning the relevant industrial chemicals. It's one of those strange paradoxes that an effect from global temperature rise is that winter temperatures over the poles drop well below normal, and stay that way for longer. The lower the temperature there, the more effectively ozone is destroyed by natural atmospheric chemistry. Temperatures in the upper reaches of this zone, around 30 miles (50 km), can drop well below their normal −170 °C in late winter when no sunlight reaches that region. As soon as the sunlight returns, the ozone is replenished. The year 2006 is said to be the date when the Antarctic ozone hole is on the way to mending, but other reports later in the same year claim to the contrasy.

During daytime in summer the normal temperature actually rises from around −55 to about 10 °C, or similar to sea level in temperate climates. Two natural features prevail in the Southern continents that don't in the North.

1. Higher sea temperatures stimulate sea life to metabolize lots of naturally volatile chlorine, bromine, and iodine compounds known to destroy ozone.
2. Mt Erebus, at around 2.5 miles (4 km) high, is one of the more active volcanoes (Figure 3.9). It can spew ozone-depleting gases into the stratosphere, the critical atmospheric zone which can drop to comparable height of Mt Erebus in midwinter. Noxious gases aren't ejected way up to these layers as in the temperate regions but simply dumped directly into the stratosphere.

Extensive cloud cover most of the time hides the effects of these from ground view and will make an interesting long-term research project. It lowers ozone levels worldwide and we all end up with a skin glow.

4.5 Mesosphere

This is very much an understudied region; this might provide some of the keys to our understanding of the fine-tuning going on in the climate.

As you go upward through this layer the temperature gradient returns to familiarity and drops steadily with height. Even in the height of summer the upper reaches drop below our critical −170 °C barrier and a lot of very strange chemistry takes place.

This is the region where mesospheric clouds in their various guises can be seen. They are seen in this at 30–50miles (50–80 km) height.

4.6 Ionosphere

Unknown by that name until the advent of radio, this 50–300-miles (80–500-km) band holds some of the most spectacular glows to be seen any clear night near the poles—the aurorae.

Radio waves are reflected back by the various layers in the ionosphere in a similar way to that of the visible light in the mirage effect (Chapter 8).

The effect arises from ions created by intense radiation from space. These ions are able to survive for long periods because the air is so thin that it takes an appreciable time for these (normally) unstable creations to meet another and recombine into a familiar element or compound. Radio waves can bounce off what is, to them, a reflective surface.

There are similarities in radio waves to visible light, but as their glows cannot be seen with the eye, it won't be taken further in this book.

The extremely low density at these altitudes leads to an interesting paradox. To a thermometer shielded from direct sunlight, the temperature would appear to be close to absolute zero. But the real temperature is a staggering 2,200° or more at the top. This arises because temperature is defined in terms of how energetic (fast) the air particles are, and these are whizzing about very fast indeed. As they bang into each other, some energy is lost as visible radiation and is part of the skyglow, although not the whole story.

Energetic particles from the Sun create havoc at these altitudes, and the result is a myriad of colurs depending on the energy involved, and the gas atom or molecule hit. That's why the aurorae never cease to amaze by their sheer variety of effects and colurs.

Evidence is rapidly mounting that this atmospheric layer is where much of the particulate radiation from space and the Sun is dumped. We'll have to wait to see who picks up these research topics and let us know whether or not they do have a short- as well as long-term climate effect. Carbon dioxide is relatively low in abundance here due to its higher density than the other gases in the atmosphere, so can't influence the climate.

4.7 Exosphere

Quite what effect this has on the earth below and in glow production is another imponderable, something only time will tell.

The upper boundary is space itself, which is taken to "start" here around the 300-miles (500-km) mark, where the space shuttle operates. Few, if any, of the gases found lower down occur in this layer, and the main mix of gases is simply the same as the space itself, just a bit denser. It has hydrogen brought there by solar winds, diffusing upward from the earth and some helium too, but some does arise from cosmic ray interactions with other gases.

For a very long time this was regarded as of no interest. Then someone spotted "sprites" (Chapter 13).

4.8 What to Observe

4.8.1 Clouds

A weather study has to include some cloud identification. As in another old saying, "Know your enemy," and we can go quite a long way in predicting a glow's type and whether (pun) it's seen.

Clouds in our normal weather layer have been categorized into a bewildering number of names. We only need to consider the highest above about a mile and half (2500 m) as good news for glow viewing and lump the rest together as normally bad news, considerably simpler to remember than the list of main types in the Figures 4.1 and 4.2.

The only exception to the " 'Bad-news' Rule" for clouds in the regions close to the ground occurs when the Sun breaks through gaps and we get rainbows and similar phenomena.

When the cloud layer reaches ground level where you are, it's sometimes difficult to decide whether or not to correctly describe this as a fog. However, a true cloud can often hold extremely large water droplets or ice crystals unlike a true fog; the glows vary accordingly.

I live on a small island where there is total disruption and confusion to transport links when the low cloud mixes with fog on the hills and sea mist comes in from the other direction! That's the time to pay special attention to all the intriguing mist and fog phenomena, not just fogbows.

4.9 Ground Level Clouds

These are the ones we experience doing a huge damage to our viewing conditions because ground level can be anything from sea level to modest mountaintops a couple of miles (2,500 m) high—a height at which most aircraft cabins are pressurized before severe breathing problems kick in.

4.9.1 Cumulonimbus

When these are around you simply don't want to be caught out high on an exposed mountaintop or a small boat at sea. These are thunderclouds. View them safely from a distance to see lightning and allied glows (Chapter 13).

4.9.2 Other Clouds

These are so complex to describe that we really only need to identify them if they're relevant as we come to each glow in turn (Figures 4.3 and 4.4).

Let's pass on quickly to another topic that is vitally important in our appreciation: our eyesight and what we need to know to get the best out of what we have.

Figure 4.3. *Persistent halo* An impressive halo lasted for the morning and was followed by days of bad, windy weather in May 2003.Contrails to and out of Paris and America add to the effect. See Figure 11.10 for the lunar counterpart.

Figure 4.4. *Cirrus cloud cover* Often this scene is followed by wet and wind.

CHAPTER FIVE

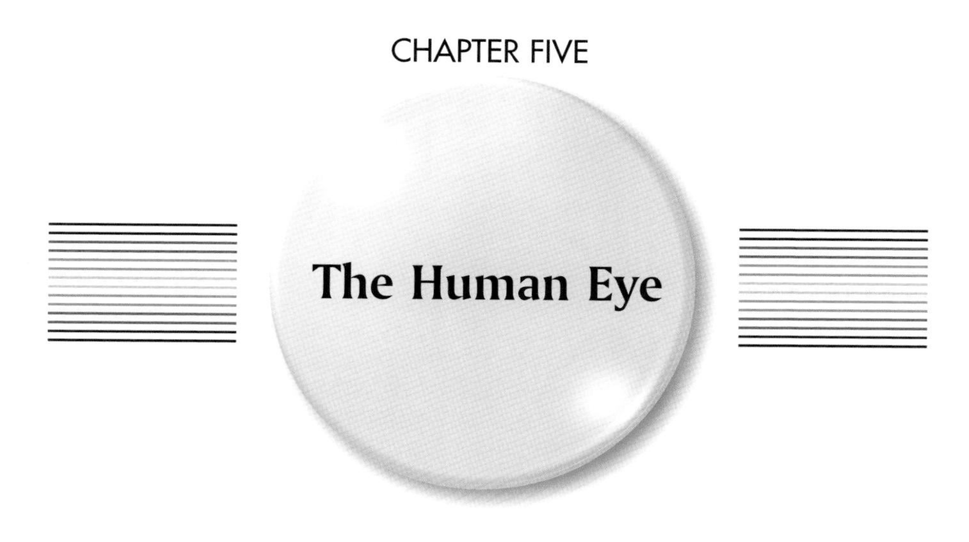

The Human Eye

No discussion on what you see and how to go about it will be complete without some consideration of the human eye. Indeed, there's no point in any of this if you cannot see properly.

Many of the unexplained sightings of strange lights in the sky can arise from an eye defect or related problem.

Vision defects fall into two categories. The more obvious one comes from some genetic or age-related feature or an accident to the eye itself. The second is much more difficult to define other than as an "unusual" bit of hardwiring to or within the brain itself where the interpretation takes place.

5.1 Mechanism

In all its gruesome detail the eyeball holds a lot of fascination, and it's only in relatively recent times that the finer working became apparent. For now all we need to consider are the three essentials—the eye's lens system, sensing surface, the retina and the "bit in between" (Figure 5.1).

- The cornea is our essential window to the world and can serve more than a single lifetime due to organ donor transplants. This window is also protecting:
- The lens itself, that incredibly tough structure, yet flexible enough to allow the muscles to pull into the right shape for focus. In good health we can expect a lifetime without trouble.

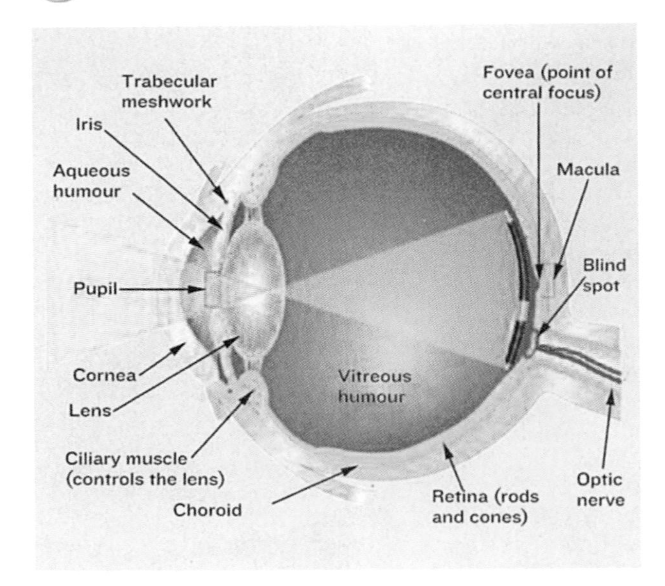

Figure 5.1. *Eye diagram* A schematic explaining how you see this schematic.

In bright daylight this is rarely a problem since the iris in the pupil contracts in size and the stopping-down (as in a camera) vastly improves depth of focus, and with it resolution. Small objects spring into sharp focus.

- The aqueous humor behind lens can have debris floating around in it. These are the floaters that often cause concern, but not always. I've given a very useful observing tip below.
- The retina at night needs to be as sensitive as possible and the iris dilates to its maximum size for allowing maximum light in. The downside is that depth of focus and acuity are severely degraded, and with it the color vision in total darkness. Where these effects apply, they are reviewed in the appropriate chapter.

5.2 Problems: "I Have Few"

5.2.1 Age

The normal aging process is familiar to all of us, as the lens becomes less elastic and viewing close up more difficult. Fortunately for our topic, viewing glows is a feature of long sight (hypermetroptric) and for anyone born shortsighted (myopic) this should be easier with age! That, however, is probably the only good thing to happen, as one gets older since a number of other eye deficiencies become more critical.

5.2.2 Cataract

Cataract is perhaps the best-known problem where the lens (often with some of the other transparent parts associated with it) becomes opaque. Diabetics will already be aware of this potential problem, but there are many other causes.

Cataract is not always appreciated as a problem until quite serious because the brain can adapt quite well. (A simple self-test is described below.) While viewing some of the more subtle glows, the serious "reddening" a cataract produces can completely obliterate the blue colors. See Figure 1.2, where blue light is now completely absorbed, one reason why the eye in this condition appears gray.

It is said that Monet's sunset pictures arose from this medical condition as an operation returned his paintings to a former coloring and brilliance.

In a layman's perspective, a cataract in the lens itself is much more serious since that couldn't be replaced until fairly recently. Those that have experienced the "op" recount a dramatic change in their view of the world, literally. This arises from the new lens's ability to transmit well beyond the blue part of the spectrum, often well into the ultraviolet and see colors we can't. In that sense they are lucky.

5.2.3 Astigmatism

Perhaps the subtler problem viewing diffuse glows arises from genetic defects in the lens and cornea combination. My own experience serves well in illustration.

My right-eye pupil is blue, whereas the left is part brown as well. The right eye when young had perfect vision, accommodating from nose-tip to infinity with ease. But it got tired quickly when looking through a telescope, microscope, or rifle sight. So I always use the left eye whenever possible with a telescope or camera viewfinder, switching to the right for finest definition. Why? Well, that left eye has a quite pronounced astigmatism, presumably arising from the mixed pupil color and genetics.

An astigmatism occurs where the cornea (sometimes the whole eyeball, or retina) is not perfectly spherical and is called cylindrical aberration. The curvature varies taking somewhat a lemon shape.

I mention this for a very simple reason. Few of you reading this now with *uncorrected* astigmatism may fully appreciate the dramatic effect(s) this can have on our perception of the world, specifically diffuse objects like our glows. When young, the brain is able to program very well and the binocular vision doesn't suffer, at least too much. As I've got older two quite worrying problems have become more acute and worrying.

The first is a "blind" area close to my peripheral vision (which is actually extremely good). By this I mean that years of my brain programming have accepted that that region is not as "important" as straight ahead. It is literally programmed out and simply doesn't "exist," and I frequently knock against things that I cannot see although within my clear vision. Judging distances as you barge in on objects can be quite embarrassing, and because of the inherent dangers I've never learnt to drive. I accept that, and perhaps I'm one of the few honest people who fully appreciate a major cause of road accidents—"I just didn't see him" being all too common an excuse.

I make a lot of this feature because reports of strange objects and the like in the sky might have a partial explanation from an uncorrected astigmatism. As something swims into the central vision it can suddenly appear, as if by magic.

The other critical feature with astigmatism is the way an image is blurred, particularly colors. Although sharp-edged objects can be seen well, such as this typeface, the inevitable assumption is that the rest of visual acuity is correct. Absolutely that's not so.

As I've got older I can no longer pick out some of the familiar diffuse glows or nebulae. The Orion Nebula used to be a very easy object but because the light area is smeared, it simply vanishes until I put on correcting spectacles (not needed at other times as I've mentioned for reading) or use binoculars. The converse phenomenon puzzles many who have not experienced it. I can still pick out other fainter glows because they are made up from a myriad of stars, which are point sources and don't blur nor smear. M44 or Praesepe is an easy object for me but the brighter M42 Orion Nebula seems to have vanished - a cautionary tale or a good explanation why we can be deceived and some things can be seen by some and never by others.

5.2.4 Color Blindness

Potentially a serious problem for about a third of the male population, the topic is reviewed better with your optician, in any good medical textbook or the web.

5.2.5 Side Vision

A much underrelated topic is given some airing below.

5.3 Remedial Action

The eye is our literal window into the world and it always pays to get an eye test at regular intervals so that one can be forewarned of a host of subtle effects. More specifically to pick up cataracts and high pressure in the eyeball at an early stage before color vision can degrade to a dangerous level in some occupations.

If you suffer from any eye problem, never hesitate to consult a qualified optician. The corrections now available are quite amazing and I hesitate to recommend laser surgery without an independent second opinion. Much safer to stick with what you have and apply a replaceable (prosthesis) corrector, or even a complete lens replacement. Take your pick from these more conventional possibilities: normal glasses, regular or varifocal, or contact lens equivalents. Modern progressives offer some enhanced properties once fully accommodated.

A full discussion on how the retina operates and adapts to varying lighting conditions is something for you and your optician. All we need to know now is that it works and what you have is to be preserved at all costs. We all have our blind spots, don't we?

The eye is such an important factor in how we see our lights in the sky that we must get that right before we can step into the fascinating world of observation.

5.4 What to Observe

5.4.1 A Sensible Self-test

I'm greatly indebted to John Wall and pass on his advice given to Steven James, my optician at that time:

Spotting visual defects
A method of self-examination of internal eye defects.
Visual observers who are curious (or worried) about the state of their eyes can clearly see defects by selecting a low-power orthoscopic eyepiece of around 40 mm focal length and viewing a distant bright streetlight, not more than 100 m away. All floaters and other detritus will be seen clearly and, more importantly, if there is developing cataract, this condition is motionless and obvious.
I use this method to monitor periodically my own developing cataracts. Floaters, which move around, look alarming but are of no consequence. Almost all of us have them.
J. Wall

How many of us, I wonder, have ever considered what effect these floaters in our eyeball's liquid can have on our eyesight in terms of sudden flashes of light and similar experiences? Steven James was certainly of the opinion that they can degrade the finest detail when using high-powered optics. I'm convinced for myself that many of the sudden flashes I see arise from my excellent side vision picking up reflections from bright lights in my (brain-programmed) blind area. How many of you have had similar experiences?

5.4.2 It Flashed Before My Eyes

Another possible explanation for sudden flashes of light only came to light, so as to speak, when the first astronauts went into space. They often complained of these flashes and got quite worried for a time.

The explanation turned out to be due to Czerenkov radiation from highly energetic cosmic rays. As these particles travel close to the speed of light, they seem to disobey one of Einstein's famous rules that nothing can travel faster than the speed of light. Not true as these particles end up traveling faster than the speed of light in the eyeball's liquid. The upshot is that light, which is to say Czerenkov radiation in our case, is emitted, which what the astronauts saw. A similar explanation is offered as a brain stimulus by the same energetic cosmic rays, another quite worrying factor as we contemplate voyages to Mars.

Seeing Czerenkov radiation is on the borderline of investigation in this book because it does require extremely good optical telescopes and equipment to see this at all. However, there are some experiments well under way on La Palma in the Canary Islands, yielding much information on cosmic ray showers (Figure 10.3).

I have seen this glow for myself, as have lots of others. It's the familiar bluish halo seen in nuclear reactors and the holding ponds for spent fuel (Figure 10.2). This shouldn't be a glow in the sky we consider before Chapter 10.3, I simply mention it now as the same phenomenon as some serious astronomy.

5.4.3 Visual Acuity

Before finally leaving a consideration of all our lens problems, it is always necessary to add in the factor of light levels and visual acuity, the ability to see small, in our case, point and diffuse sources.

Staring straight ahead is what we do most of the time, but that situates the eye's dead spot right where it focuses. That's never a problem in practice because our eyes constantly move, we blink, and so on. However, it does mean that the eyes can get tired in bright light conditions, and more so in the dark.

What do we do about this problem when trying to spot some of the fainter glows? Very simple really: you don't look at them at all!

That strange piece of advice is really based on our primeval survival needs. Most dangers appeared from the side out of the bushes, and the eye has become much better than you realize at detecting movement and faint objects in its peripheral vision. To see some of the fainter glows more distinctly, in day as well as in night, *never* stare straight at them. Always look slightly to one side.This is called "averted vision."

The best position to look is some 8–16° off center. Tilt your nose that much and see what happens.If you've never tried this trick I can assure you to expect a pleasant surprise, because it works. Every time. Once learnt, never to be forgotten (Figure 5.2).

Figure 5.2. *Montana eclipse, 1979* Another example of a persistent halo leading to weeks of freezing weather in United Kingdom, a third of world away. This picture is to be used to test your averted vision, or how much more apparent is the faint halo. It also shows contrail shadows that caused so much discussion at the eclipse in 1999.

Another interesting feature of our eyes is also worth checking, that of our color vision at the periphery. For this you need a helper or something that will hold two cards of different color some distance apart.

Move to and fro and stop when you are aware of the two cards at the edge of your vision. Even at my advanced age I can still see colors but most of you reading this will find it difficult or impossible, which is more normal. That phenomenon occurs because the eye color sensors or cones are bunched at the center of the retina. The rods at the edges are needed for brightness, and movement sensors are needed to spot danger.

Not having color sensitivity at the corner vision is often a major hazard in our age of garish warning signs. Fortunately, the slightest eye or body movement instantly detects the novel light and, provided the clue is taken on board, all is well. Otherwise, an accident statistic joins the multitude before it and hence the following tale.

5.5 A Salutary Tale

Using averted vision is extremely helpful but the eye can still play tricks. I pass on this gem I picked up in 1998 from the 8 August British Medical Journal. Mark Pettigrew from York wrote this salutary tale.

Uranus attacks! Herschel's legacy

Towards the end of the eighteenth century, the astronomer William Herschel also provoked consternation by observing that a large comet was heading in our direction. He confirmed this with a series of measurements over subsequent weeks showing it increasing in size: from about two seconds of arc on 17 March 1781, to three seconds on 19 March, to four seconds a few days after that, and so on. Clearly something was on its way, but, unlike the asteroid earlier this year (Toutatis), it never arrived. In Herschel's case this was because he had actually discovered the planet Uranus, and it was not approaching the earth, but moving away. Yet Herschel's measurements clearly show the size increasing, and instrument error could not explain the systematic increase in size, night after night. Historian Norris Hetherington has suggested a simple answer: Herschel's strong prior beliefs had overridden his natural objectivity, and his measurements reflected this. Despite careful calculations Herschel simply succeeded in supporting his prior hypothesis that the blob in his eyepiece was a comet.

Herschel's legacy, therefore—apart from discovering the planet Uranus—is a reminder that humans are inherently biased in recording or measuring events, even those as objective as the movement of planets. We are easily led astray by our natural inclinations to see what we want. Naturally this failing does not apply just to astronomers: Schulz and colleagues have pointed out that controlled studies that are not double blinded tend to overestimate effect sizes by 17%. Although this is perhaps not as earth shattering as reporting that a comet is heading towards the earth, it is a useful reminder that as scientists we just cannot help kidding ourselves.

The moral of this old tale is extremely clear. Never attempt to make accurate measurements of all the glows discussed in this book unless you understand the problems involved and are aware of the many tricks the eye and the mind can play on you.

Figure 5.3. *Safer viewing by reflections* The safer view of the sun through tinted copying to enhance the effect when printed here. No optical effects this time around the sun.

So, I repeat. The result is that all of us really do need to get our eyes checked over at regular intervals. It can save our eyesight if a problem is diagnosed in time.

5.6 A Final Warning

The book's intention is not to scare the hell out of you, but more to encourage the beginner to go out and have fun.

However, we must pay strict attention to safety and preserve our eyesight, and where necessary I've repeated that homily throughout. Here's one way of looking at the Sun with a degree of confidence (Figure 5.3).

Now we can relax and read on and better appreciate with a happier glow.

CHAPTER SIX

The Twilights

Long before humans became truly sentient, they fully understood the importance of twilight.

In truth, that instinctive bit of "hard-wiring" in the brain did mean the difference between life and death and has to be a prime, if not the prime, reason we are here today. Our eyes are very well adapted to seeing under low light conditions, even by the stars and nightglow alone on a moonless night as we'll read later, but that's still not quite good enough to spot everything nasty with fangs that crept up on us from the shadows.

Our ancestors introduced the term "twilight glow" into our language, making this the first well known of our lights in the sky. For that reason alone, we are justified in placing it on top of our list.

Dawn heralded the birth of a new day and with it a temporary freedom from the unseen perils of the night from savage predators.

- In urban environments, maybe do we still need these instincts and abilities?

For most of us in the twenty-first century the transition from night to day is nothing more than a signal to sleep or work. When it gets dark, we simply switch on a light and carry on with our chores until we decide to sleep. It was never so until humans learnt to control fire and extend their working (and social) skills' period with the flames lighting up their caves. Dangerous animals were also deterred from approaching too closely (see Chapter 2).

Other animals also knew that the time day turned into night was very dangerous for them and took the necessary precautions to hide. All animals seem to be aware that once the first glimmer of light or glow appears on the horizon, the new day is on its way and the transition to full daylight is simply a question of waiting as it gets brighter.

Modern man has decided that the transition isn't that simple. The "Men in Grey suits" do seem to get everywhere.

There are no less than six twilight zones in the course of a full day, three at each end.

Throughout this book I'm regarding the morning twilight as a much more important transition under the blanket title of *dawn's aglow*, and that's why I will only consider that example in detail. For the *dusk* twilight sequence the timings go in reversed order.

6.1 Mechanism

In defining the way the Sun moves below the horizon we obviously cannot "see" it there. Astronomers always mean the "mean" point at any time and for us that's wherever the *center* of the Sun happens to be.

Another complication we don't need to worry about is the way the apparent size of the Sun varies during the year, as we don't orbit it in a circle, but an ellipse. The definitions below refer to the center point of the Sun (to get over any and all of the other complications) as that point is always the same whatever size the Sun appears to us. It's not the "bit at the top" as you might expect.

- Because the Sun is about half a degree across, this can lead to some funny paradoxes when living close to the poles. The average person thinks of sunrise as occurring when the first sliver of Sun appears above the horizon, but, as we shall see, that can appear to be early due to mirage and other curious atmospheric effects.

For quite a long time on the way toward spring at the beginning of the year, the "official" sunrise (and then toward fall at the end of each year, sunset) times will not match what you see. The Sun will skim along the horizon with a visible upper portion on view, yet recorded as not rising (or setting).

- For all our considerations here the twilights' time bands really only apply from the equator to something short of 50° for all the year.
- From this 50° region to the Arctic and Antarctic circles some of the twilights extend throughout the whole night as midsummer approaches.
- In polar regions, special factors apply with permanent night, twilight, and daytime in turn for extended periods. What constitutes a true twilight does not apply for a considerable period on either side of midwinter.

Twilights are defined to start and end when the center point of the Sun is that many degrees below the true horizon (let's assume a sea view).

6.2 Astronomical Twilight

It is much more important from our point of view if we wish to stand any chance of seeing the very faintest lights in the sky.

Let's define this as the true boundary between night and day. Stars at the very limit of visibility to normal eyesight at magnitude 6 start to vanish as the sky shows definite glow at the horizon.

This is the real beginning of the ancients' true twilight and has, thus, always been known. Modern man calls this "astronomical twilight."

At dawn the start point is tied down to be *defined* as the time when the Sun reaches a point 18° below the horizon. It's this "legal" definition that leads to our many paradoxes as we get closer to the poles.

6.3 Nautical Twilight

This can be regarded as the first of our true twilight glows.

This is the transition period from just making or blundering your way about in predator-infested undergrowth (side streets), full of the deepest shadows of night, into enough incipient daylight to switching off our modern artificial torchlight's stage.

The history of its introduction is obvious from the name and signifies a particularly hazardous time for navigation when the horizon interface between sea and sky is fraught with false sightings—or rather a surfeit of them.

Nautical twilight's start is *defined* as the time the Sun reaches a point 12° below the horizon.

6.4 Civil Twilight

From our earliest sentient times this was the only twilight that mattered. It's the time when the glow has increased to a point where things can be seen clearly enough to work without some form of artificial light.

Civil twilight is *defined* to start when the Sun has reached a point 6° below the horizon.

6.5 What to Observe

For all that have been defined above, the timings are taken from the center of the Sun's disk.

Twilight effectively ends in the morning as soon as only half the Sun is visible over the horizon. That's what leads to our curious paradox that in certain parts of the world on a particular day in spring and fall as the Sun just skims along the horizon, twilight is all day! True daylight never occurs.

These definitions are parceled up in this neat 'thirds' bands of (6) angular degrees below the horizon way for convenience, but as everyone knows life's not that simple.

- At the tropics, these divisions make a lot of sense as the angle the Sun cuts the horizon is more or less constant throughout the year. The timing of the transitions remains extremely predictable, with sunrise always not far from 06.00.
- Twilight is at its shortest everywhere at the two equinoxes.
- Anywhere in midlatitudes, which we take to be from the tropics of Cancer and Capricorn to 40°N and S, the timings are valid and we can observe the twilights at very precise timings.
- Things start to get right out of hand as the Sun rises closer to the crucial 18° below the horizon limit as summer approaches.

London in the United Kingdom and Winnipeg in Canada at 50°N are just above that cutoff point (there are no major cities in the South). Places on this line never leave astronomical twilight all night long with the Sun skimming under the Northern horizon at this critical 18°. That period is some 6 weeks long.

Table 6.1. A Table of twilight timings for critical dates in the year

	March 21		June 21		December 21	
Place	Start	End	Start	End	Start	End
Anchorage	03.27	20.50	*_*	*_*	06.18	17.38
61° 10′ N	04.25	19.52	*_*	*_*	07.12	08.12
	05.17	19.00	*_*	*_*	08.12	15.44
	06.00	18.16	02.21	21.42	09.14	14.42
Alderney	04.13	04.52	*_*	*_*	05.56	18.00
49° 44′ N	04.52	19.24	02.03	22.00	06.35	17.21
	05.30	18.46	03.08	20.56	07.16	16.40
	06.02	18.14	03.52	20.12	07.54	16.01
Los Angeles	04.40	19.36	03.04	20.59	05.32	18.24
34° 00′ N	05.09	19.06	03.44	20.20	06.02	17.54
	05.38	18.37	04.20	19.44	06.33	17.22
	06.03	18.12	04.29	19.14	07.01	16.55
Guantanamo	04.50	19.25	03.57	20.07	05.12	18.44
20° 09′ N	05.16	18.59	03.57	19.36	05.12	18.17
	05.42	18.33	04.57	19.07	06.06	17.49
	06.04	18.11	05.21	18.42	06.31	17.25
Mt Marapi	04.55	19.19	04.43	19.20	04.39	19.17
0° 20′ S	05.19	18.55	05.10	18.54	05.05	18.51
	05.43	18.31	05.36	18.28	05.31	18.25
	06.04	18.11	05.58	18.05	05.54	18.02

- The further north you go the more the twilight bands vanish until we get to the Arctic Circle where the Sun famously never sets in summer.

Although not a rigid rule, it can be said that these regions above 50° north and south to the Arctic or Antarctic Circles are the best places to see many of the upper atmosphere glows because the sunlight penetrates these regions. On the ground the darkness is almost total as we see and experience it with adapted eyesight.

The year 2006 turned out to be a bumper year for sightings well below that 50°N latitude limit.

- Just in case nobody's risen in time, twilight is seen as a segment of a circle of ever-increasing bluish light centered above the Sun's point below the horizon.

This point will move to the right in the north, but always vertically above the Sun. In the western sky the dark patch of true night starts to lighten and move down in synchrony. Most people see the night sky region as purple black, whilst others describe it as indigo-blue.

As a newcomer to glow watching, what's your description of the appearance and colors at both sides of the sky? Check with Section 7.6 of Chapter 7 for some answers.

Is there any obvious link between the two hemispheres and the bit immediately above you, the zenith?

CHAPTER SEVEN

Dawn Glows

7.1 Introduction

Dawn glows are so much less popular and more is the pity, I strongly recommend the reader to concentrate on dawn glows for two extremely simple and practical reasons.

- The eye is already fully dark adapted as far as it can go, something guaranteed if you stay up all night. The less palatable alternative is to get up early enough, dress, then go out into the dark for at least 15 min. Proper dark adaptation is genuinely necessary to get the best views in a predawn period.

  The really faint glows can be savored without further effort on your part. As one gets older, dark adapstion takes longer and finally gives up almost completely when cataracts and the like start to be a bane. In the dusk glows, you're relying entirely on the eye adapting fast enough to cope with the rapidly changing and diminishing conditions, and where subtle colors are involved, these are usually totally missed.

  At dawn the increasing light level and color acuity becomes a positive feature, allowing each glow to be followed a lot better in real time as the eye becomes more color sensitive.
- As dawn breaks few people are awake to see the many glows that herald the new day.

Contrary to expectations, this has to be a major virtue as some glows are much improved when seen then. The air hasn't warmed up enough to create interference from heat haze and heat mirage (Chapter 8) and light pollution is markedly much less with more people asleep than in the evening.

This may often be the only practical time to view in urban areas when domestic light levels are minimal. The only drawback brought to my notice is a higher tendency to get arrested for loitering!

I'll now deal with these dawn glows in much more depth and cross-reference to the dusk when necessary.

The exception has to be with the only really inevitable, and therefore truly natural, glows. These are the twilights, a genuinely mirror situation. They're covered in Chapter 6.

It is always worthwhile looking at a dawn sky in case phenomena more commonly associated with the evening or true night sky pop into view.

This can be a fitting reward when staying up all night to see the true night astronomy topics. That's why we start in more or less chronological, rather than alphabetical, order with Zodiacal light and its associated Gegenschein. Both of these extremely faint lights in the sky are more commonly considered a nighttime feature, requiring extremely dark conditions and perfect skies.

Some comets can only be seen in a dawn sky, of course.

Almost any of the night or day glows could be included in the category "Rare Glows at Dawn", together with most of those seen at any time.

That leaves one other genuinely of the dawn.

7.2 Zodiacal Light

This dawn glow will forever remain a mystery to all living in an urban area. The prerequisites for any view at all must be a crystal clear sky and no light pollution.

Another reason why this is rarely seen is that it's seasonal, even in countries over huge tracts of North America, Europe, and much of Asia. The best time for it is the full months of September and October, when the glow makes its sharpest angle to the horizon and skies at their cleanest just afore dawn. That's the main reason for rare sightings, the early hours when few nonastronomers are awake.

We cannot regard the Gegenschein as a true dawn glow. The sky clarity conditions are much more demanding and a time around midnight is best to see this. The seasonal factor also has to be taken into account, but if you live in the tropics, always make an attempt to see it all year round just before dawn to make this another dawn glow.

7.2.1 Some History

The sight was so familiar in so many early civilizations that it's impossible to say who first identified the glow. We can accept that the first scientific document probably lies with Aristotle who used the word "zodiac" to describe the band of sky that the Sun appeared to traverse in the course of the year. The literal translation comes from the 12 constellations as "Circle of Animals" although quite how Libra fits into this scheme beats me. On the science proper, Aristotle was completely off beam and attributed the thing to gases from within the earth, such as volcanic, very familiar in that part of the world as Sicily was within Greek influence at the time. The gases somehow yielded all the "misty" glows. We know today that volcanic effluvia diminish our glows through atmospheric pollution, but might have an effect on the Zodiacal light coloration.

The Zodiacal light has featured in many pieces of folklaw and some extremely important history. Perhaps the best known is the Andean cultures that based their constellations on bright regions of the sky like the Milky Way. The Zodiacal light was well integrated into this pantheon, adopted by Montezuma II further north in Mexico.

The Aztec Empire was also guided by prophecies and high on the list was the return of Quetzalcoatl on a raft, heralded by the appearance of the right sights in the sky. A particularly bright Zodiacal light was seen in 1519, followed rapidly by two comets, and what he heard as voices. When Cortez arrived immediately afterward, Montezuma handed over his empire without a fight to what he assumed was the god, who'd been seen to arrive by ship.

The king must have known about Zodiacal light being a fairly common sight in that part of the world, so that particular apparition must have been unusual in some aspect or other. We now know that the brightness and extent can vary tremendously and in concert with solar activity. If there were a particularly savage solar storm in 1519, it also might have been accompanied by electrical discharges in the atmosphere, when crackling noises are not unknown. Aurorae at the tropics are extremely rare but also not unknown. We'll never know for sure as European astronomy had yet to pass the Inquisition into modern science. Chinese records might help.

7.2.2 Mechanism

In theory the Zodiacal light should be best seen in midwinter when the skies are darkest but the geometry is wrong and the ecliptic merely skirts the horizon where atmospheric absorption and pollution is at its worst. Another negative factor coming into sighting it then is very simple; the brightest part of the Milky Way starts to appear in the dawn sky closest to sunrise. If you do live on a rare mountaintop in a northern continent with exceptional skies, you might just be able to spot the glow at dawn and chalk up a rarity. For most, this will be an impossible time of year.

- The summer is also a very bad time the further north you go because of the ever-increasing twilight zone with the ecliptic running, for all intents and purposes, horizontal to the ground.
- The best time for late-risers to see this glow is the converse situation in the evening in spring.
- The further north you go, the less likely you are to see anything it would seem but when in polar regions the air is almost totally devoid of moisture and that factor alone aids visibility considerably.

Those in a southern country will find the best dawn seeing conditions in their fall, our spring months of March and April.

The way the ecliptic cuts into the horizon determines where and when it is best to view. For our morning glow the full is the best and for the Southern climes the spring. For those lucky enough to live in tropical places near the equator, the Zodiacal light can be see at any time of year as sunrise (and sunset) occurs at roughly the same time each day and the ecliptic always cuts the horizon at a sharp angle.

For a long time the cause of the Zodiacal light and Gegenschein remained a mystery, only becoming totally obvious when satellites explored regions well outside the Moon's orbit. It arises from sunlight reflected back from the myriad of small particles orbiting the Sun in our ecliptic plane; particles that either never made it to coalesce into planets proper, or debris arising from their collisions, or both, plus a huge amount derived from comet tails as they venture close to the Sun.

The larger fragments could well be large enough to be regarded as extremely small asteroids, but the bulk is microscopic or considerably smaller in size. There will also be considerable gas and chemicals around in this background as well as water in the form of ice. Depending on the amount of ice and its size and crystalline shape, this can and will have a dramatic effect on the brilliance.

This cosmic debris is in constant motion, not just in orbit around the Sun. The larger particles tend to migrate into the Sun and are consumed, whereas the solar wind travels ever outward leading to what would seem to be a diminishing source. However, comets and asteroids proper generate plenty of new material whenever they orbit close to the Sun, and it's from that origin that we know the Zodiacal light will go on forever due to replenishment.

This stuff is an extremely good light reflector and is confirmed to be the origin of the glow by a study of the spectrum. It's plain ordinary sunlight.

It also explains why the Zodiacal light exhibits its unique geometry. The splurge of light closest to the horizon is so much brighter there than anywhere else in the sky as it's really a combination of light by direct reflections together with every form of forward scattering. We see all these reflections best when closest to the Sun. To all has to be added a huge amount of ionization radiation in that intense environment. The best way to think of this region of space is an extended atmosphere of the Sun, even though much of it is particulate matter.

The geometry also indicates that the glow is from the equatorial band of the Sun.

Taking all these factors together, the brightest features will be closest to the Sun and, therefore, ideally seen immediately before dawn in the northern full. That mandates a perfectly clear sky and a mountaintop for preference. A small island with no land heating and light pollution in the way is a good substitute (Figure 7.1).

7.2.3 What to Observe

The best description is that of a cone of light poking up from the horizon with its brightest part at the horizon and markedly tapering off to invisibility. As you see it from the ground the cone is a sharp triangle tapering off rapidly.

Figure 7.1. *Zodiacal light* The brightness is greatly enhanced by a long exposure. Expect to see this at limits of averted vision. Photography as in this long exposure enhances the faint Zodaical glow at dusk. Flip the images for a "dawn" example!

- Even under the best conditions the cone rarely manifests itself more than a few degrees in extent and is blending into the general night sky blackness long before 90°.
- A normal site well away from the tropics will be lucky to see a 20° extension from the horizon.

From this brief description it's easy to understand why the Zodiacal light is so difficult to see. You must pick the best or right time of year. Then the brightest part is at the horizon where atmospheric absorption is at a maximum and the fainter portions rapidly merge into the background. A crystal clear sky has to be a must.

The most we can expect to see under our normal polluted skies is an indistinct "splurge" of light immediately preceding the Sun's first true rays. This is the brightest region of the Zodiacal light proper and that's why it shows up more often as a hemisphere or semicircle of lightening, rather than a sharp extension as I've just described it.

As normally seen, there will be band of light stretching up from the Eastern horizon at an approximate angle of 23° to the right. The ancients often described this as a false dawn.

Whenever seen it's necessary to do more than just note that fact.

- The extent across the sky is a useful guide to sky clarity.

  Perhaps I can suggest this sequence of observations as an alternative way of monitoring solar activity.
- How bright is this compared with some other familiar sight? In October the Orion nebula (M42) is ideally placed to make this assessment (Figure 10.12).
- The inner or lowest clump will have a shape and an apparent entity of its own. By how much is this a percentage of that seen and is it elongated as normal or more semicircular?
- Are there inner structures? Dark bands can be explained as clouds in silhouette, but if these don't move. . .maybe we see the structure of some uncharted meteor stream or comet.

- The final aspect is by far the most important piece of science to undertake. Does the brightness change markedly with time during the night, and/or adjacent days?

This last point is related to many historical tales of bright coloring. Montezuma is said to have seen colors, and as I've noted here there might just possibly have been aurora displays. However, that doesn't explain a much earlier independent observation by Seneca in the first century. He described the sight as "scattered fires" and likened them to distant and thereby unseen flames. There are many recent observations of strong coloration, of all colors and hues, and none can be explained away as just atmospheric extinction. There might have been localized volcanic activity in Sicily/Pompeii or Mexico, of course, adding to a temporary dawn reddening; the dusk Zodiacal light apparition would go unnoticed or be invisible.

For my money, assuming the coloration is really confirmed by modern spectroscopy, the huge variations in the human eyesight and color vision in previous times go a long way to explain many of these sightings. Finally, a bright spike is often recorded. This is highly reminiscent of the spike seen with the Arend Roland comet in 1957, a perspective effect from illumination of the dust tail.

Can any of this be linked to solar activity? Strong solar storms occur very frequently at solar sunspot minimum, as they have in the recent cycle. The minimum was supposed to be around 2005 and yet 2006 turned out to be a bumper year for magnetic storms, aurorae, and noctilucent clouds.

A final thought: the solar wind can and does affect minute dust particles in space. Several satellites have been launched to study these interactions.

7.3 Volcanic Dust

We've all seen that famous painting "The Fighting Temeraire" by Turner, and others that show some extremely dramatic dawn, daytime, and dusk coloring. This is not due to bad eyesight nor artistic license. The explosion of Tambora in 1815 had a devastating effect on the earth's atmosphere and led to 1816 being described as the "Year without a summer."

It took years for the dust to settle and in that time the sky took on a red to purple hue at all times when sunlight was visible. There have been countless occasions in earth's history when similar volcanic eruptions affected both the atmosphere's clarity and with it the climate. That's the normal downward moderator in a general upward trend to a warmer climate. All volcanoes had some effect on how the ancients saw the sky above although surprisingly few are recorded or can be identified. That lack of written commentary isn't proof on its own, but when taken with the geological records it does confirm that there have been relatively few really big events in recent centuries.

Perhaps Krakatoa in 1883 is the best described and the first to be treated to a full scientific recording. The evening and morning skies were reddened for many months. The modern eruption of Mount Pinatubo in 1991 was a sobering reminder of what can happen. A few weeks later, by July, in nearby Hawaii, that single eruption had sufficiently altered the local climate/weather to disrupt the eclipse viewing. The normally dry desert region saw rain and clouds, whilst the rainy highlands cleared enough to see some of the event.

By November the volcanic dust had covered the whole globe and led to some spectacular twilight coloration in Namibia at the tip of Southern Africa. The skies remained reddened for some months afterwards and the global temperature drop was around half a degree, the figure depending on the source of the data.

7.3.1 Mechanism

For a volcanic eruption to affect the atmospheric seeing in this global fashion it isn't sufficient for there to be a simple "bang." An explosion on its own will discharge huge amounts of material and if that's mainly boulder-sized rocks, they soon come back to earth and do no harm. Nor will gases as these rapidly disperse through the troposphere and are washed out by rainfall in a short time. Molten magma also doesn't provide any threat for the same reasons, being too little in amount to affect more than local conditions.

The real culprits are volcanoes liable to pyroclastic flows. Here the molten magma is of a particularly viscous type and contains a high amount of dissolved gases. When it blows, the pressure is released and any magma is blown into smithereens. The mixture is incredibly hot and gas laden with two results.

First, and most common, the pyroclastic flow simply tumbles down the mountainside and creates havoc wherever it goes, at speeds sometimes of hundreds of miles/kilometers an hour. For a really big one, the second situation, the amount of material ejected is so vast that the superheated gases carry dust particles well up into the stratosphere, well above the weather layer. That's why this dust can take decades to fully return to earth, and materially distort the weather patterns by blocking off sunlight. The sulfur dioxide carried up also has a considerable effect on atmospheric chemistry, which also changes the amount of sunlight and its spectral range at ground level.

The general name for all these factors is "haze." This all happened with Pinatubo and with many others in the past and is bound to happen again, soon, maybe before this book appears in print!

7.3.2 What to Observe

Whenever there's a really big eruption the localized effects are dire and a total blackout can last for days if the winds aren't very strong. These effects don't need further description, being so obvious in the way sunlight diminishes and changes character. Dawn effects become somewhat academic.

The prime condition for a spectacular dawn in the ensuing months on the other side of the globe is an otherwise completely clear atmosphere for several hundreds of kilometers to the East. Where there are clouds the sky colors cannot readily be distinguished from a normal red dawn.

- At the first tinge of twilight, the first unusual sight will be a pink to purple hue, appearing over much of the sky. Depending on the severity this will range from pink to true purple at the horizon only and then gradually inch up to fill the whole sky in a period of 10–20 min. As this glow is more normally seen at sunset that's termed a "secondary glow," although here it's a primary effect. It is also given the more obvious name of "purple light." This extremely nice glow is more than worth the effort of an early rise since the eyes are much better dark-adapted than the other end of the day. The eye doesn't need to adapt to a decreasing light level as it's already working that way.
- By the time the purple light has reached a full intensity, around the 20 min mark, a secondary (primary at sunset) glow is seen as a strong band of light along much or all the Eastern horizon. What makes this glow so characteristic and different from a normal dawn is the sheer intensity of the lighting. It's derived from reflected sunlight from the huge mass of small particles in the upper atmosphere, a region well above the normal clouds.

Depending on the size, quantity, and time since eruption, this dawn glow goes through many fascinating transitions in the course of time.

With Pinatubo, the purple coloration merged into a strong red tinge at the horizon and in the course of about half an hour this band edged ever upwards and transformed into an orange then into more of a golden color. Just before sunrise the whole sky was incredibly brightly lit by what can only be described as white color, quite unlike a normal blue sky. It was only at true sunrise that sky coloration merged into anything familiar.

What distinguishes this glow from a normal sunrise has to be the intensity of the coloration and "weirdness." Savor its sheer beauty. The picture from Namibia in 1992 shown in Figure 7.2 and used as the book's cover background picture was taken at dusk but the same sights were in view at dawn.

Anyone with complete color blindness can never fully appreciate the way the reds and blues interact to create such strong purples and the transitions from red through orange to gold.

7.3.3 What to Anticipate

Because volcanic eruptions are an inevitable feature of the earth's history, plus the fact that we don't have the vaguest idea how to predict them, we must always be on the lookout for volcanic sunrises after an event. In 2006 there have been many false alarms with Agun and Marapi in Indonesia, and the whole of the Andes chain looks set for a major event. If Yellowstone does "blow," then we'll have much more serious matters to worry about than meteorology!

There were innumerable serious ones in Indonesia relatively shortly after the 1883 Krakatoa event at the end of the nineteenth century, followed by Martinique in 2003 and Katmai in 1913.

The curious factor in all written history before the Industrial Revolution is the almost total lack of any reference to atmospheric volcanic effects. That's in spite of the undoubted influence in the Black Death and later collapse of the Iceland

Figure 7.2. *Namibia* In November 1992, a year after the Pinatubo volcanic eruption, dust affected the Southern Hemisphere. This sight was seen at dawn as well as dusk.

and Greenland populations after Laki. That leaves only one written record in the Bible and the Santorini (Thera) blast, which was considerably bigger than Krakatoa, about six times more.

This ancient volcanic eruption was one of the most catastrophic eruptions in history about 3,650 years ago. There wasn't a single eruption but a whole series over about a century but the big one shot an estimated 72 cubic miles (114 cubic km) of debris into the atmosphere, and the onslaught of dust and acid droplets would have darkened the skies and severely disturbed the climate.

We know that this almost certainly wiped out the thriving Minoan civilization nearby, also disrupting the whole of Greek culture as recorded in their sagas around the time of Troy. Historians differ on the chronology but it's now likely that the eruption may have helped the Israelites escape from Egypt. So, we ought to be prepared for a dawn starting "It gave them the guiding pillar of smoke by day and fire by night," described in the Bible.

Santorini's collapse into the sea would have sucked in a gigantic volume of water onto its hot magma creating yet more explosive force and setting off a devastating tsunami throughout the eastern Mediterranean. The situation is very much as seen in the Sumatra tsunami of 2004. It's also tempting to think that the Israelites would have experienced similar as the sea was sucked out, leaving them the seabed to cross. A tremendous wall of seawater would have followed soon after, certainly capable of wiping out the pursuing Egyptians as they chased behind.

As they say in all good books and TV programs, refer back to Figure 3.11 on these effects in detail. "Don't try this one at home." I'm lucky to be alive.

7.4 Other Dusty Glows

Having just dealt with all the doom and gloom anticipated from volcanic activity, what else is on the horizon? Unfortunately, we must expect a lot more of the same from many other dusts in the atmosphere.

This dust is always increasing and in our future may well turn out to be the most significant factor of all in determining whether or not we survive to see some, or all, of the other glows.

As reviewed in the chapter on light pollution, the contributory factor from backscatter from "stuff" in the atmosphere is by far the biggest culprit.

Country and island light-pollution refuges suffer extremely badly from external dust hazes, and the chances of seeing the faintest glows like Gegenschein are diminishing all the time.

7.4.1 Mechanism

7.4.1.1 Desertification In the absence of a severe volcanic event or cometary impact, the earth's climate will swing ever more decisively into the arid conditions it normally experiences in recent interglacial periods. Some definitive evidence for this has already been seen in the African droughts for the last four or more decades. Figure 3.2 of three decades ago demonstrates on a very small scale how dust reaches the upper atmosphere. Figure 3.4 shows that problem worsening.

Much more significant are the 2006 crop failures in huge tracts of central Asia, specifically in the "Grain Basket" of China and Mongolia. Northern Afghanistan has had (effectively) a total crop failure.

It's ironic that much of the fertile land in those regions is derived from the very same source, which is windblown dust deposited in great depth in a relatively short space of time. We even get the name of "loess" from this phenomenon. The vast fertile region in China and much of Europe was shaped like this. My own small island has a lot of it, several meters thick overlying the base granite.

Each spring, for well over half a century now, clouds of fine yellow dust sweep in from the Gobi Desert and northern China's arid lands. After decades of overgrazing, deforestation, and heavy water extraction the desert regions grow ever larger with sand dunes edging Beijing. Today's best estimates indicate up to a third of China has become desert or is turning into one. The glaciers feeding the rivers have largely disappeared and by the end of the century neither will be left, turning the whole region into a parched land with no plants binding the soil. The great dustbowl in central United States in the Depression period will pale to insignificance in comparison with this permanent ecological switch.

The spring of 2006 also saw the worst dust storms for several years blotting out daylight in Beijing and cloaking it and other cities in choking clouds of dust. Millions of people gasped for breath, overwhelming hospitals with respiratory emergencies.

This was not a localized problem in 2006 and dust often reached Korea and Japan, where it closed schools and airports, and factories had to take emergency measures to protect goods being covered in dust.

Nor should we be complacent on the other side of the globe. In the French Alps in 1990 skiers discovered yellowish brown snow in the dawn. Two weeks earlier a dust storm in this same Taklimakan Desert region in China had blown more than 20,000 km across the Pacific and then the Atlantic. Europe also regularly gets the same treatment from the Sahara, and motorists will often find a yellow to brown stain on their cars before off to work in the dawn.

Although not widely appreciated, one of the commonest reasons why the Canary Islands' observatories have to stop work is this same dust problem from the Sahara. Whilst the dust remains in the upper atmosphere, it "merely" degrades any study of the faintest glows; when it arrives at mountaintop height the places must be covered completely to save the delicate mirrors and mechanisms from abrasive dust damage (refer again to Figure 3.3).

Is this dust man-made? I don't think so as the process has been going on since time began, and there can be no doubt that greenhouse gases are not the root cause. That said, the overgrazing and destruction of habitat in the Middle East in the last few millennia has had a substantial part to play in the expansion of the Sahara. Simply we're seeing more of the same in Asia, and increasingly in the Americas.

Unless there's a marked switch to wetter climatic conditions the dust source from deserts can only get worse.

Unfortunately that dust source isn't the only one to worry about.

7.4.1.2 Deforestation A major cause of desertification lies in the way mankind is consuming the forests. Not just in the Amazon, but everywhere. Local deserts spring up immediately, and more critically the microclimate is radically altered, and any flash flood sweeps the topsoil away. That may or may not affect the weather locally and hence climate worldwide. But that's only the political part of the story.

California has always experienced forest fires that are part of the natural ecosystem to ensure sequoias regenerate. The worrying factor is the increasing number of other places experiencing huge fires that never did before. The single case that brought this to our public attention occurred in Indonesia on 19 September 1997. Raging fires in an equatorial forest created so much smoke, our second type of dust, that the authorities in several countries a long way away declared a state of emergency. This closed schools and offices in Sarawak in Malaysia for 10 days, thus not a minor event. See Figure 3.6 for "hot" news as this book goes to press on that on-going problem.

Six months later at the other side of the Pacific another vast fire in Mexico caused similar havoc over most of Central America. There have been many similar tragic losses of forest since then and 2006 has seen some of the worst episodes in European history throughout Iberia.

At the time of writing this, at the end of August 2006, thousands in Greece were being evacuated in a state of emergency.

The astronomical effects are twofold. A recent fire in Siberia added no less than 2% to the carbon dioxide tally into the atmosphere, "fueling" the debate as to whether or not this constituted man-made or natural causes to global warming. More seriously, the smoke acts very similarly to volcanic and desert dust in the destruction of our views of the fainter glows.

7.4.1.3 Air Pollution Until the Clean Air Acts came into force in most industrialized nations, air pollution wasn't an issue. People just accepted it as a fact of life. The London "Pea Souper" is etched into my memory, with sights of men preceding the trams with lit braziers, little more than an arm's length in front. They were already vanishing into the murk in that short distance.

The only glow we ever saw for days on end was a very strange one—that of a dim red object in the sky. That was the Sun. Forget ever seeing anything as bright as the Moon.

Whilst it's hoped those days will never return, the sheer number of vehicles and other sources of particulates means that concern has never entirely gone away. In the West, we enjoy Los Angeles smog at worst (Figure 2.4), and as seen in Figure 3.5, but China is reverting to those days, and cities regularly disappear into a dense atmospheric haze of industrial effluent.

All this adds to the world's total audit of atmospheric contaminants and degrades even further our enjoyment of the natural glows in space.

7.4.1.4 What to Observe The drastic effects from these other atmospheric dusts all combine to make our artificial dawn glow very similar to that from the entirely natural volcanic source. Expect to see an increasing frequency of purple to red glows in what is otherwise an apparently clear sky. On the downside, never again expect to see any of the difficult fainter glows in the night and at dawn.

- Desert dust behaves very similarly to volcanic material because it is, for all practical purposes, composed of the same stuff. Expect the effects to be more seasonal, particularly in the Asian spring, and not lasting more than a day or so. This material is wafted airborne by weather pattern winds and will not reach a great height. Rain and gravity rapidly clean the air, whereas volcanic material can stay in the stratosphere for up to a decade.
- Wood smoke from forest fires has a much more subtle effect but can also stay within the lower atmosphere for a surprisingly long period. That's because the particulate matter isn't the whole story and some very interesting compounds indeed join the mix, or are formed later by solar action.

  Like desert dust, the effects don't remain entirely local and can straddle the globe, and the real origin is never identified. The mess just adds to the worsening background.

  Because the particles and molecules are so much smaller, the sunlight is considerably more reddened and blues largely eliminated. Expect to see effects at dawn similar to a view through a bonfire or a stubble fire when driving along a motorway.
- True atmospheric pollution is the real villain of the piece since it's with us the whole time, in ever-increasing amounts. On a normal windy day the stuff gets wafted away to add to the background. Most is washed out of the atmosphere

very rapidly although the pure chemical constituents can go on to do other things, a topic out of the scope of this book.

You don't need a temperature inversion as in Los Angeles to see the grim extent of this "micro" smog. It's quite easy to spot after a period of relatively low wind over any densely inhabited region as a purple to brown haze a few tens or hundreds of meters thick.

The effects on atmospheric clarity are extremely dire and some strange lunar coloration was quite common in the heat of August 2004 (Figure 3.5). Major cities will always be prone to vehicle pollution and Athens seems to suffer from this all times of the year.

Expect to see similar effects at dawn as with wood smoke.

- Perhaps the only truly natural haze comes from plant effluents. The gum trees in Australia are the best known and that haze can hang around in valleys for much of the summer.

All these dusts can dramatically affect or rapidly change weather patterns as the dust seeds rain clouds. Always keep a "weather eye open" and anticipate one or more of the more specific glows caused by them. At least, in that sense, some minor good will comes out of it.

Whilst taking every precaution, you can gain a very good measure of how critically bad your site or atmosphere is by carrying out the "Bishop's Ring Test."

If this passes, you are an extremely lucky person these days. Faint glows remain on the menu.

7.5 "Red Sky in Morning"

"Shepherd's Warning" is how the traditional jingle goes. And there's quite a lot of truth in it in the Northern Hemisphere.

7.5.1 Mechanism

The prevailing winds come from the Southwest bringing the right type of clouds to the Eastern horizon; they are already with you for impending rain and storms to follow very soon afterward (Figures 4.1 and 4.2).

As we face north in this schematic diagram, cirrus clouds high in the sky presage the approaching warm front from the left, which is from the West to southwest. At a normal rate of progress in a clear night these will be seen in the morning sky just disappearing over the Eastern horizon.

7.5.2 Sunrise

- As the first rays of the Sun become visible in the twilight they shine through the whole length and depth of these wispy clouds, becoming reddened in the process.

- That red light shines onto the base of more of the following cirrus and cirrostratus, illuminating them and adding to an overall hue by reflected light.
- If the conditions are right, this red light will shine onto the next band of cirrocumulous, which are often the characteristic fluffy red blotches over much of the eastern side of the sky. This part of the sky is not normally brightly illuminated in a clear dawn.

You know you're in for trouble if the whole sky becomes brightly lit red with denser altocumulus and stratocumulous clouds, and big trouble with a red western cloud base on the other side of the sky.

Under normal conditions the warm front will pass over fairly quickly and is probably the origin of another piece of familiar folklaw "Rain at 7 will cease afore 11."

This glow heralding a wet morning is often the first and only thought an average commuter gives to the environment about them. It's good start, but no more. The weather it portends really does put a dampener on the prospect of seeing anything until later in the afternoon, unless you get the good fortune to see lightning or a rainbow as the front begins to clear.

- A red sky in morning shouldn't be taken too literally when it's little more than a mild reddening as that's bound to happen whenever any higher altitude thin cloud cover is over the horizon.
- More often than not the following altostratus arrives too soon and extinguishes all red long before seen or noticed by the late riser. That does mean listening to weather forecasts as rain follows very shortly afterward.

7.5.3 Moonrise

"The Moon in haloes hides her head"

Because the "red sky in morning" saying is so well known, few bother to take more heed than that. A red sky and rain is on its way; no rain and all stay happy.

By far the better indicator of a rainy day on the way is another bit of folklaw in the "Halos" jingle. Anyone awake before dawn will notice a moonrise with a bright halo around it. This is *not* a corona (Chapter 8), but a true halo.

- The approaching warm front will push the highest cirrus clouds over the Eastern horizon and as these contain ice crystals of the right type, light is refracted through them to give these pretty halos.

  The halo cannot be seen at a normal sunrise because the Sun's light is much too bright when well up and by the time that has happened, the more dense clouds blot it out anyway.

This much better indicator of imminent wet weather is seen when the Moon is surrounded by a thin ring of bright light, of the same sort of diameter as a solar halo.

Because of that familiar optical "delusion," the Moon always seems much larger when near to the horizon, and the halo can appear to be enormous. In reality, it is of the same size as seen with a noonday Sun. There may well be other arcs as with solar phenomena but rarely noticed due to the much lower light level.

The characteristic of this wet harbinger is a white ring of light with a reddish inner tinge. Hence it's a true Red dawn.

7.5.4 Evening

The reverse situation of a red sunset isn't due to the same cause except in the rare situation that a front passes over later in the day. The reddening under those conditions can arise from illumination of the retreating clouds by the setting Sun.

7.5.5 Other Causes

There are three other major reasons why a red dawn can occur, although in reality they are the same. These are dusts in the atmosphere from either a dust storm, forest fire, or volcanic eruption of some magnitude somewhere, often the other side of the world. Seeing those effects at dawn is much rarer due to late rising.

7.5.6 What to Observe

Very little really to observe other than checking these different variants and identifying their validity.

A really good red sky seen by you is always a good topic of conversation to "brighten up" the dismal day with your dripping colleagues as they arrive at work.

7.6 Earth Shadow

Whether or not this can be considered a true glow is a difficult one. I've only included it here with dawn glows because it sits naturally in a real-time sequence with the other dawn glows described in this chapter.

It's entirely due to an effect that allows it to be seen when particulates and dust are in the atmosphere. Strictly speaking, it's what the name describes—it's a shadow, but to see it you must have a glow backdrop.

7.6.1 Mechanism

The earth's shadow may be seen anywhere and at any time there's a twilight glow. Some criteria in an approximate order of relevance make this sighting much more likely.

- A clear sky and a perfect (flat) horizon in the East. Gaps in an eastern cloud cover rarely allow enough sunlight through to show this to full effect.

- A murky atmosphere to act as some backdrop in the West. That can be natural cloud, dust, or man-made pollution.
- A moderately flat horizon all around. Low hills shouldn't be a problem, particularly to the north or south.
- A high vantage point helps considerably. Being on a mountain or in an aircraft, you should always look out for the effect as most of the bad weather clouds might, or at least ought to be below you, but not affect sunlight through higher layers between them and you.

  The tops of skyscrapers make unusual vantage points in cities where a background of pollution makes a sighting almost a certainty if the sky is otherwise clear enough.
- A view around the winter solstice in the Northern Hemisphere.

  The summer solstice also has the Sun rising much more slowly than at the equinoxes but grossly extended twilight isn't as much of a problem as knowing when to make a very early rising. Neither factor should deter attempts.

7.6.2 What to Observe

When the Western sky is totally clean and clear, the chances of seeing anything are extremely thin. A shadow can only be cast onto something and natural high cirrus cloud works extremely well.

- Just before the true sunrise is the best time to see the shadow appearing as a dark circular patch extending a few degrees up across the whole of the Western horizon. The correct time is when the sky is lit enough to see the contrast effect. This is quite obvious as a shadow of the circular globe of the earth thrown onto the backdrop.

  The color seen depends to a huge amount on what's in the atmosphere to the East and the material forming the Western screen. Expect a blue to grey hue seen against the normal pink.
- Under good conditions there should be a clearly defined reddish upper border to the main shadow that appears to merge into the diffuse pink sky. The width of each band varies considerably according to sky conditions and time of the year.

  Even with poor color discrimination in your eyesight the contrast between light and shade still makes this an easy object.
- As the Sun progresses above the Eastern horizon the shadow moves rapidly downwards, appearing to move a lot faster than the sunrise. It finally flattens out to match the true horizon and vanishes in a small fraction of time just before the first sliver of sunrise.

  That distinctive sequence of features confirms that what you've seen is the genuine one of earth's shadow and not a cloud shadow effect.

Only count this in if you wish to include every possible dawn (or dusk, Chapter 9) event.

7.7 Crepuscular Rays

Of all the dawn glows, these rays cause the most confusion in their description and terminology. Correctly defined, crepuscular rays come into the category of shadow effects and appear as dark bands across the sky. However, in order to see them, there obviously must be a backdrop of light and that's where the confusion arises.

7.7.1 Mechanism

If the sky is more or less evenly illuminated, we have true crepuscular effects. The bands are quite obvious and that's all there is to see in an otherwise normal dawn sky.

The confusion arises because it's much more common to (mentally) ignore the dark bands and to only see other rays that are considerably brighter than the background illumination, superimposed upon that background. These are "sunrays" caused by sunlight coming through gaps in whatever is causing the shadow effects, normally clouds.

Because seeing a bright phenomenon is so much easier than the lack of it, we can regard sunrays as the better description for our purposes simply because they fit our definition of an obvious "glow". It all adds to the fun in life (Figure 7.3).

None of these effects can ever be seen when the air is perfectly clean and dry. They all arise through light scattered back from atmospheric particulates. At dawn these tend to be at a minimum, hence the relative rarity of sightings when compared with dusk.

7.7.2 True Crepuscular Rays

These are only seen when the Sun is below the horizon.

Figure 7.3. *Sunrays and crepuscular rays* Photographed while writing this book.

Anything seen during true daytime cannot be because of a very simple piece of geometry. A true crepuscular ray stretches from horizon to horizon, something that cannot happen when the Sun is up and in part of the sky. A shadow can only be shed from wherever the Sun is at that time to the horizon below. It's also extremely difficult to see shadows against the full daytime glare, and that's why most daytime sightings attributed to crepuscular rays are reported during or shortly after a rainstorm in clear atmospheres when some rain and mist remain in the air, or shortly after dawn when pollution is rife.

Figures 9.1 and 9.2 show the effects very well from desert dust in the air, at dusk, when these are more frequently "noted".

7.7.3 What to Observe

At any time when there's a lot of "stuff" in the atmosphere expect to see any of the shadow effects, and when there's a lot of broken cloud close to the rising Sun, crepuscular rays become extremely likely.

- Look out for tight bands of darkness against the twilight's coloring. These bands show up most clearly where the light is strongest and get more indistinct as they go overhead, to peter out somewhere before you fall over in following them. In really good examples, these rays should be seen to go overhead and then pass through onto the opposites side of the sky when they should become more distinct again.

  You know these are true crepusculars when the bands stay tight all the way and as they go through to the opposite horizon appear to converge again. This convergence point is the antisolar point, which is to say immediately opposite wherever the Sun is at that time. With true crepuscular rays this point lies in the counter-twilight, which is a deep purple color in a normal cloudless sunrise. Most people see this as black, which is why shadows are so difficult to see there.
- You shouldn't see sunrays at the same time. If you do, these will show up very bright and be quite obvious against the twilight background. Sunrays invariably peter out long before they get anywhere near to the zenith. They are always the more spectacular phenomenon and will be talked about long after the dark bands have been forgotten.

  Points to note here include the obvious factor that you will be looking at a very confused sky that changes extremely rapidly. There are at least three types of bands involved in this case; crepuscular rays, brighter bands due to the background twilight (or countertwilight), and sunrays.

  When the atmosphere is extremely clean and the sunrays are seen against a backdrop of natural water vapor, then these shafts of light will go from an intense bright yellow to golden color. Quite obvious and always spectacular. The more the other material present, the sunrays, although still bright in comparison with the twilight background, verge into the orange, through crimson into ever-drabber reddish umber. Whilst this is not true science, keeping a record of the number and colors seen and for how long they last might be of some use to future historians monitoring the climate.

- Although of no scientific value, it is an interesting exercise to calculate in your mind where the clouds are over the horizon causing the shadows. Immediately at sunrise, that distance should be quite obvious as you can see the cloud formations and the points of origin. When the antisolar point is any elevation at all, you have the necessary angles to work the mathematics for the distance over the horizon. Expect a huge surprise as these distances can lie between 90 and 380 miles (150–600 km).

  When the bands appear wider apart from the anticipated sunrise point; the calculations can track back from there, particularly from photographs taken at the time. Your elevation determines the distance to the true horizon.

  An alternative way to calculate the cloud's distance over the horizon is to time when seen before sunrise. In northern temperate countries, a degree in longitude is 70 miles (100 km to make the maths easier) and that's 4 min in time. That means true crepuscular rays should show up around a quarter of an hour before sunrise. Time both and you can make very accurate distance measurements from the real longitude spacing.
- Do the bands appear to diverge, then reconverge at the antisolar point? This is an entirely optical effect due to the eye/brain's need to make sense of what's seen.

  If you dare to stand in the middle of a straight road whilst making this observation, does that road appear to do the same? Compare your observation with another road lit by lamps at regular intervals. Safer still, use telegraph or power lines which are visible in the sky.

We can now make a "beeline" to the next dawn glow.

7.8 Solar Pillars

Extremely rare dawn sightings make this glow of particular interest. The phenomenon is caused in the same way as so many others by ice crystals almost directly in line of sight with the Sun. Because the air conditions tend to be more stable at dawn, sightings ought to be more likely, not less so.

The other fascinating feature of solar pillars is that there isn't a single type but a whole spectrum. The shape and alignment of the ice crystals determines everything.

7.8.1 Mechanism

7.8.1.1 Crystal Types The mathematical calculations that one can do for all the possible snowflake shapes and sizes is mind-boggling. There are simply too many possibilities and you'd never get to the full number even in a raging blizzard. That said, all the myriad of flake geometry boil down to two basic ice-crystal shapes: rods and sheets (Figure 11.1). Stuck together in endless permutations is the basic way they come into being.

How these two basic forms fit together is the key to it all but in practice irrelevant in the creation of a pillar glow. The prerequisite for their appearance is not ice crystal diversity but their conformity and regularity. The pillar seen depends on most or all the crystals being of a type, then those being aligned in the same way. If the crystals are completely randomized, reflections off the faces cancel out and the space close to the Sun appears its normal whiteout.

That basic requirement doesn't preclude the appearance of other arcs and halos, so always keep a keen eye open to spot the more elusive pillars which are usually ignored when the more brilliant sights appear in daylight. Solar pillars only become the main features immediately before sunrise and after sunset.

7.8.1.2 Sheets The best way of describing this fundamental crystal shape is as a hexagonal tablemat (Figure 11.1). The thickness is extremely less compared with the width.

Light can pass through the sheet with a change in path, of major importance in other glows, but not here. For a pillar to be seen, the sunlight has to be reflected off the flat surface and the first instinct is to expect a majority of these to be aligned more or less vertically. But it doesn't work quite that way.

In a normal ice crystal cloud the natural tendency is for crystals to fall that way but in stable conditions they can settle horizontally. When this happens we see sunlight reflected off the bottom surfaces, very much as from a ceiling at the end of a long corridor. Added to this reflection will be an almost equal intensity from the upper surface of the crystals, which is why the effect is so striking. The light pillar will be bright for several solar diameters upward and not much wider than the Sun itself, tapering off rapidly at the top as the reflection angle directs the beam away from our eyes.

In full daylight it is quite possible to get a downward solar pillar from these same crystals creating the dawn sighting as the Sun rises higher. These descriptions swap places as the Sun passes the zenith to make complementary explanations for the sightings in the evening skies and at dusk.

7.8.1.3 Rods Ice in this crystal shape (Figure 11.1) is also quite common and is best described as a tiny hexagonal sheet that's grown in the opposite way and created a pencil. The length is considerably longer than the end diameter.

As with sheets, the necessity for our pillar sighting is pencils lined up, this time they can be horizontal or upright. They must all point in the same direction, which is across our line of view. This is a much rarer event because, on the one hand, the reflective area is so much smaller and, on the other, only about a third is correctly pointing at any one time. In a normal cloud the crystals might well be spinning on their long axis, and this is actually of no importance since the effects will be to "smear" out the reflective effect and render it more stable.

As with the plates, the pillars seen with rods will appear above and below the Sun in daylight.

7.8.1.4 What to Observe Although I have seen solar pillars many times on land, the necessary weather conditions do seem to be much rarer there. What we need are steady or stable atmospheric layers and these tend to be upset by thermals and a rapidly changing weather pattern. The classic pictures in history all seem to come from cold climates like Antarctica, where the ice crystals form readily and wind flow is constant. These same conditions apply at the coast, and solar pillar sightings from Alderney are extremely common and can be attributed to the airflow being in stable layers. This is laminar flow where the air isn't forced to rise over ground at a different temperature and create turbulent conditions in the process.

Expect to see solar pillars whenever there's been a period of stable weather and the upper atmosphere shows signs of cirrus or other "wispy" cloud formations.

- Whenever a true solar pillar appears above the horizon at dawn, expect to see other arc and halo sights during the day. This by no means a golden rule, merely a hint of things to come. The uncertainty arises from the way the weather patterns arrive in the Northern Hemisphere and it's quite likely to be a system departing as one arriving.

  An excellent example of this turned up after I'd written this chapter, and that series of sightings is described in the last chapter.
- The pillar's color and brightness gives many clues to the mechanism leading to the sighting. Really bright ones tend to be caused by plates, and stable weather patterns allowing the crystals to align.
- White pillars do not always foretell good weather to follow much more the possibility that the atmosphere is relatively clean.
- Strongly tinted pillars can mean almost any weather pattern to follow as the coloring is inevitably due to a lot of absorption in the air between the Sun and the crystal cloud formation and/or us. If the sky is brightly red lit as with a "red sky in morning," that's part of the same weather pattern. Expect bad weather to follow.
- A solar pillar with crepuscular and Sun rays is worth an early rise because of the rarity.

  The Sun isn't the only bright object liable to show pillars.
- These are surprisingly common whenever the Moon is above the horizon before sunrise. The lighting is at its best then and with a perfectly dark or nearly so sky, the pillars show up a lot clearer.
- The brightest planet Venus is well known to show pillars and if you ever see this planet through hazy clouds, record anything seen.
- Jupiter is certainly bright enough to exhibit halos and this planet is another to check on regularly.
- Whenever there's a fog and that's a freezing one, always check the appearance of a bright light some distance away before it disappears into the murk. Halos are normal and when in absolutely steady air, bright pillars can appear too.

All these bright objects may show pillars above or below, or both together. Modern video equipment is easily capable of recording these faint glows. Set the controls at maximum sensitivity for night scenes.

Do not use a filter in front of the lens, and always use manual control to focus on these illusive features (Figures 7.4 and 7.5).

I hope you will try this sometime.

(A)

(B)

(C)

Figure 7.4. *Solar pillar* A sequence showing how a solar pillar developed immediately before Figure 7.3.

(D)

(E)

Figure 7.4. *(Continued)*

Figure 7.5. *Late solar pillar, France, 2006* Another dawn pillar, easily missed in the background glare. Always check similar scenes by averted vision.

7.9 Green Flash

At one time considered entirely an optical "delusion," seeing a genuine green flash at dawn is something to savor. The scientific basis for the green flash had to wait until 1960 when D.K.J. Q'Connell of the Vatican Observatory produced the first authenticated color photographs of the phenomenon, but he took those at sunset, when sightings are most common. Monochrome films taken by others over the years couldn't confirm a color.

7.9.1 Mechanism

The green flash occurs when light is refracted through an atmospheric layer and thereby seems to come from this higher layer. There may be several layers, of course, each or all sometimes giving a separate episode.

The green flash's origin in refraction from air layers, usually of a differing temperature, makes this a special case of *mirage* (Chapter 8), normally associated with a hot-daytime sighting.

The characteristic coloring arises from a combination of factors. Red light is largely unaffected by atmospheric absorption and passes straight the way through so that you see the Sun as reddened. Blue light is scattered completely and never reaches the observer's eye. That leaves the bulk of sunlight, which is predominantly yellow to green to pass through the refracting layer and give a distorted image of the first vestiges of the Sun's upper rim.

Another major consideration in green flash formation is the rate of sunrise. At the equator the Sun rises most rapidly and whilst sky clarity and mirage are at their best, a flash lasting more than a second or so is rare.

The converse applies at the poles when a temperature inversion coincides with the Sun's disk appearing to skim along the horizon. In the months around the summer solstice just inside the Arctic Circle the Sun performs this trick and a green flash can last, it seems, forever.

7.9.2 What To Observe

For a dawn sighting a considerable amount of guesswork is involved and that does make the sighting worthwhile. However, there are some very important criteria to anticipate before even thinking of seeing one at dawn. Timing is definitely not the first of your worries.

- Unlike the situation at sunset the Sun's shape and color cannot be seen and used as a predictor. The main factor in favor of a dawn sighting is the much more stable air then, the prime requirement. The more stable the atmospheric layers, the merrier. An inversion layer also assists, or may be an essential component in the process.

  At dusk you can see the circular solar disk settle into a weird series of distinct slices and know that things look promising. A frosty night is often your only predictor for a dawn apparition, particularly favorable with ground frost rather than a general air frost.

- The next vital predictor is an absolutely clean atmosphere. More than a trace of dust or other pollutants and the subtle coloration(s) will be missed.
- The final vital factor is an absolutely genuine horizon. Whilst some flashes can be seen when the whole solar disk is well up, they are so rare as to be regarded as a bonus, not the main theme. Airline pilots are in the best conditions and regularly report flashes and tend to regard them as normal, not a special thing. They have an unrestricted horizon with the lower atmospheric layers well below them and when flying East see everything speeded up. Few aircraft allow for backward viewing when going West! For pilots the best views occur at dusk when flying West so that sunset is slowed up almost completely.

  The next best situation has to be from a mountaintop overlooking a vast plain. The classic viewing point is one looking East from a shoreline. These days, many on a cruise liner deck more or less guarantee sighting some, sometime during the voyage.

  The higher the deck on a liner, the better and it can be great fun running down the decks to catch the sunrise a bit later. This stratagem makes a lot of sense the further north you go as sunrise is amazingly prolonged in summer as the Sun skims along the horizon. Alaska and Scandinavia are now regular cruise centers.
- Binoculars or a telescope is not vital but considerably assists rapid identification when the flash lasts only a second or less.

Safety is paramount.

In the more northern climes, eyesight protection becomes a major concern with so little pollution and atmospheric water present to cut down the full glare. This is another reason why dawn sightings are a safer option with much less temptation to prolong viewing when the bright Sun pops over the horizon. In the evening, too many observers start much too early when the Sun is well up and try to follow it down.

Whenever there's any doubt that the Sun is too bright for comfort, leave all recording of the phenomenon to a video camera, where there's no direct eyesight involved. A burnt-out camera is no real loss whereas one's sight is a catastrophe.

Always have a very dark pair of sunglasses in hand, or the correct solar filters used for eclipses.

This is one occasion when solar projection through an optical device onto a screen is unlikely to be satisfactory, although the only genuinely safe mode.

Once you've settled into a cruise, plan to rise early and see the flash.

- A genuine green flash will appear just above the horizon as a horizontal line or smudge, a very short time before the upper rim of the Sun appears. In some cases the two will appear attached; in others a marked distance will separate. This is the exciting feature; you never know what to expect.
- Provided the Sun's brightness is not too great, continue to watch the solar disk and plot the shape. When this breaks up into slices, a series of green flashes

becomes quite possible as each layer emerges or moves up from the horizon. For an explanation, see Chapter 8.

My record is in seeing nine separate flashes as the layers go through the refracting layers at the horizon. But that was at sunset.

- The actual color seen might not be uniform and can vary a lot from bright yellow through patches of green through to an almost uniform blue. Blues should never be possible and it is difficult to give an explanation.
- A detached red flash is also seen on occasions. The explanation here is a bit easier when seen at the lower edge of the solar disk when fully risen and is due to the red light refracted and reflected back along the layer responsible for the first and rising green sighting. This red band is always much easier to spot when there is some cloud cover obscuring the main solar disk. It isn't a contrast effect and can be photographed (Figure 7.6).
- The real gem in any series of sightings is a fully risen Sun with a greenish cap, which may or may not be fully detached.

  The Sun isn't the only bright object in the morning sky to exhibit the green flash.
- Always keep an eye on the moonrise and chalk this up if seen. The overall light is so much lower that it's a real test of visual acuity.
- Venus is a potentially interesting subject and has been recorded with a green flash.
- Jupiter is bright enough to be included in any survey.

Knowing exactly where they rise is the second challenge and a telescope becomes essential. In really clear skies the planetary disk will be seen rising without too much difficulty.

A shore site in the tropics is the place to start, as a ship is too unstable with the telescopic magnification needed. See this rarity with the naked eye if you can, of course.

Figure 7.6. *Ottawa, 1970* At dawn a detached green flash exhibited by mirage (Chapter 8) due to freezing ground in winter.

7.10 Specter of the Brocken

This glow can be seen at any time of the day or night and is normally thought of as one at dawn. The prime requirement is some mist in the air or dew or hoarfrost on the ground, all of which rarely persist till dusk. Regard it as special to this time of the day.

Strictly speaking the effect is best called the "specter" and is yet another shadow effect. The "Brocken" is caused by interference due to light passing round some obstruction, usually yourself. When the glow is seen around your head, then this is the true and full-blown effect. It is often referred to as "The Brocken specter."

7.10.1 Mechanism

Whenever there's a bright light behind you, a shadow is cast. In clear air this is never seen at head height and is normally only picked up as a very familiar dark shape on the ground. During the night a shadow in front of you is much easier to see when there's some mist or fog about, then the eye–brain combination starts to play tricks.

Without some clear distance marker, obscured by the mist, the eye sees this shadow as unusually large. This effect is magnified in the brain because the shadow isn't in a single place as on the floor in daytime, but cast for some distance *into* the mist. The human eye is notoriously unable to focus on that variety of subject form or anything without a firm edge.

It's also a lot darker than expected. Coming across this specter when passing in front of a single, bright light source at night can be quite alarming. This phenomenon is a true specter.

At dawn the Sun's light is intense after night adaption and is more frequently seen then.

With a very bright and small enough light such as a rising Sun, some light will be diffracted around your shadow and might be seen as a faint halo only around head height. This is where one's eyes are, of course! This is the genuine point opposite to your eyes in the sky from the light source. In dawn light that's the real antisolar point.

Most of the light won't come from this diffraction at all and will be reflections back from the myriad of small water droplets (or ice crystals when extremely cold). This makes it simply another one of the halo phenomena.

The Brocken part of the name arises from the mountain in Germany where it was first recorded and by far the easiest way to see this is looking down from a high point with mist at ground level or in a valley.

7.10.2 What to Observe

The fun and game in seeing this glow come from the huge variety of places and times when it's possible to see one.

- Spot a specter. Any time, particularly in a deserted wood at night.

- Anywhere in a city with a single spotlight behind you; that isn't "Jack the Ripper" waiting for you.
- From a high spot with mist in the valley just at sunrise. Try to guess how far you think the shadow stretches into the distance and how wide is your halo. Then see how many colored bands you can trace out.

Immediately around the head will be a dark purple circle that then merges into a green tinge, which turns red at the outer rim. Under excellent conditions there will be another outer ring of light, almost as bright with the same color sequence.

Under perfect conditions two or more concentric bands will appear outside the inner two.

Always carry a camera with you and capture your "saintly" image (Figures 7.7 and 7.8) spread out on fresh morning dew instead of as others might think of

Figure 7.7. *Brocken* Any time there's dew on the ground, look out for a specter of the Brocken.

Figure 7.8. *Brocken* The dew can be suspended on tall grass, and then "halo" to a Brocken.

you—as a specter. The question then arises—is this a true Brocken or our next dawn glow, Heiligenschein?

7.11 Heiligenschein

Literally translated as "Holy Light," this is often confused with the Brocken because they appear very similar. A genuine Heiligenschein needs much more specific conditions. The mechanism is quite different.

7.11.1 Mechanism

Like the Brocken and the specter this can rarely be seen other than at dawn. The basic difference is that you don't need a mist or fog *in the air* to reflect light back into your eyes.

- You must have dew-or frost-covered ground or foliage. In this case it's extremely easy to judge the distance because grass and other familiar objects are in your field of view. The shadow seen is sharp and distinct simply because that's entirely normal.
  The basic difference from a Brocen is the way light is reflected back.
- It's not from the minute water or ice droplets themselves, but from the minute solar image focused onto the surface of whatever they sit on.
  And that requires yet another important specific condition.
- The droplets mustn't be in contact with each other, in effect to create a film, but well spaced and, at the same time, close to each other.
- They mustn't be in contact with the surface but some distance above it, and that's normally found with "hairy" foliage. Grass isn't always like that but some species are, or have ribbed leaves that act similarly or are waxy, thus stopping minute water droplets from making direct surface contact.

Sunlight passes through the minute droplets acting as mini-magnifying-glasses and comes to a focus a short distance behind. This focus is less than a quarter of the droplet diameter outside it, and that's why conditions are so rare. The droplets, or a very high percentage of them must be held in some way that quarter-droplet-diameter distance from the surface.

You can check this situation with a glass marble, and see that the focus isn't a true point but good enough.

Each minute spot of light is extremely intensely bright, the burning glass effect, one reason why dew evaporates quickly and why an early rise on a clear day is so essential.

Each intense spot of light is then reflected back through the same droplet into your eye, but as the spot is so small some interference effects can also take place and that generates a colored halo in very favorable cases. All in all, not a simple mechanism operating but a series of all those specifics must happen.

Then there is another final and vital condition.

- The sky mustn't be cloudy or hazy, rather a nice clear blue sky. Any fuzziness around the Sun making that expand into a ring of glare stops the minute magnifying glasses coming to a sharp focus. The end result is then just a brightly reflected background to your shadow. Situation is normal, in fact.

7.11.2 What to Observe

From the question posed at the end of the last Chapter, Figures 7.7 and 7.8 are much more the type of thing described in Mechanism that we need to observe. Much more common in any crisp dawn and easier to see.

- As the primary condition is dew or hoarfrost and a clear sky, at least clarity around the Sun, you must get up early to check this glow.
- Because this glow only arises from reflections back from the antisolar point it's normally seen as a white halo around your head.

  You can have a great deal of fun placing a camera on something amusing and activating remote shutter release. The halo will center on the silhouette centered on the camera lens. That will prove the effect to be a real one and not an optical one in your eye.
- Coloring around the bright ring is rare. The more interesting feature is the apparent ring size and whether this can be related to drop size. Generally speaking, the droplet size isn't important.
- The full gamut of ground glows is extremely complex and outside the scope of a discussion on glows in the skies. The complications arise because the reflections aren't always from and through the droplets. Dew bows can occur because of reflections within drops only and are true rainbows. The eye can play funny tricks, as the sight seen might not be an arc of a circle but other geometric shapes.

  Take pictures and see what geometric shapes you can make out, then identify them.
- These colored reflections can occur from almost any surface covered with small and closely packed discrete objects. Debris floating on water is a favorite.
- A final place to look out for is provided by the modern reflective clothing worn by emergency personnel or others wishing to be seen at night. The minute glass beads on this material make excellent reflectors for our purpose and a sheet or coat spread out on the ground will provide hours of endless amusement. Similar are some of the reflective road surface markings.

One of the best I've seen comes from a low Sun reflected from road signs. The colored halos and intensity of coloring are much better seen if the sign's background is white, and that can be seen from some considerable distance away.

All these pretty effects are incredibly nice to see and, as with street signs, quite unexpected.

7.12 Nacreous Clouds

Very much a rarity and unlikely to be seen by anyone outside Polar Regions, these are at their best at the South Pole. They mustn't be confused with noctilucent clouds (Chapter 10), nor with iridescent clouds (Chapter 8).

Because of their appearance they are also called "Mother of Pearl" clouds. They come into a category of meteorological phenomena only found in the stratosphere, and then only in this restricted geographic area. The generic name for them is "Polar Stratospheric Clouds" (PSCs).

They are included in a dawn sequence because they are best seen in a short hour or so period before dawn (or similarly after dusk). In Polar Regions this means they are only seen in winter, quite unlike noctilucent clouds only seen at the other end of the year in summer.

7.12.1 Mechanism

These clouds form well above the normal weather in the troposphere, starting around 3 miles (5 km) further up into the stratosphere, to about a third into that higher layer, or 9–16 miles (15–25 km). This places them well above our normal commercial aircraft flight paths over the Poles from United States to Europe and Asia.

- They are not formed directly, rather from something going on way down below at ground level. The mechanism is one of a pressure wave set up as strong winds pass over mountains, as the wind speed is never exactly constant. The wind flow creates alternating regions of low and higher pressure, and as the partial vacuum appears downwind on the lee side, minute ice crystals are formed. This is precisely how a Wilson Cloud Chamber operates, hence that name from the same concept.

  The main difference from the Wilson effect is the temperature, amazingly low at around −85 °C. At this height the air pressure and temperature combine their effects so that the water vapor doesn't condense as fog, but as extremely uniformly sized ice crystals, typically 10 μm. This uniformity of size sets up ideal conditions for sunlight's diffraction and interference effects.
- Pressure waves over mountains need very strong winds for the effects to persist and even more so for the pressure waves to manifest themselves as these nacreous clouds many miles/kilometers above, in another totally different atmospheric layer. Katabatic winds in the Polar Regions fit the bill (Figure 7.9).

7.12.2 What to Observe

It is not generally realized that pressure wave clouds can persist for a very long time and be seen many tens or hundreds of miles/kilometers downstream. Their stability is assisted by the extreme cold up there plus winter's poor sunlight.

Figure 7.9. *Etna and pressure waves* In more temperate climes we see pressure waves quite often, as in this example seen over Etna.

- Nacreous clouds are seen as a spectacular display of "Mother of Pearl" colors and striations in wispy clouds moving in gentle arches and curves. They rarely fill the sky and at first sight look more like classic "mare's tails" in Technicolor.
- The way these shapes form and move across the sky gives some indication of the extreme winds and conditions at these heights. These jet streams were entirely unsuspected until causing so much grief to American WWII pilots over the Pacific. Jet streams extend way above those (now routine modern commercial) cruising heights. Understanding how these streams vary and affect climate has become a major research topic (see below).

Your only normal chance of seeing them is during a Great Circle flight over the poles. Look out above you. If you see anything up there veiling the sky above 50,000 ft, this might be the genuine article, but don't expect it to be illuminated except on rare occasions in the polar twilight in winter.

Any nacreous cloud you see that doesn't quite fit the description might well be a variant.

7.12.3 Some Science

Nacreous clouds are but one type of Polar Stratospheric Clouds, Type II in fact, yet they are turning out to be some potential keys to our understanding of the mechanisms affecting the climate.

Type I are much less spectacular dawn glows and subdivided further into a, b, and c. All three subtypes are not plain water ice, rather a complex cocktail of atmospheric acids, nitric and sulfuric, for the most part. These are entirely natural, of course, but considerably added to by cars and electricity generation from coal.

All the Type Is form at a similar height and may well have our nacreous clouds with or even in them. This is unusual as Type I forms in slightly warmer conditions around −78 °C.

Ordinary nacreous clouds, composed as they are of ordinary ice, merely act as an indicator of the general weather patterns in the lower stratosphere. This is a useful function in itself.

In the case of the Type Is, the study has become deadly serious.

It's now realized that these clouds are main chemical factories involved in much of the recycling of these atmospheric acids.

In the days before industry became so universal, the nitric acid formed by lightning reaching this height was very less, and the role it plays in destruction of ozone kept in balance. This nitric acid held on ice crystals converted gaseous chemicals from natural sources composed of chlorine, bromine, and iodine (the halogens) into destructive fragments of the original molecule. These fragments (known as free radicals) convert the halogen chemicals into something that immediately destroys the ozone. The cycle repeats many times (called a chain reaction, but not a nuclear explosion this time). This process really gets under way as sunlight returns in spring.

Paradoxically, the polar winters are getting colder with the rise in summer's warmer conditions globally, and the build-up of these harmful acids is aided considerably. That's yet another cause of the ozone hole expansion, which is paralleled by the increasing sightings of our PSCs.

Today we have much artificial halogen-containing gases going up into the stratosphere, and the naturally produced varieties from increased algae and plankton activity in polar oceans, too, as the climate gets warmer. Both contribute to the now familiar "ozone hole."

At the time of writing this, new reports suggested that the "hole" is decreasing in size. However, the very weekend of posting this text to the Publisher, NASA published a report for September 2006 showing the largest Antarctica hole ever. Was this a result of the Montserrat eruption earlier in the year that went unrecorded in reports on global warming? Sea temperature's drop globally was reported as a "mystery."

That may mean the sightings of these gorgeous PSCs might decrease in the future, or they can increase hugely from the next big volcano as happened with Pinatubo. This is one occasion when we should rejoice at the loss of a nice glow.

7.13 Comets

By far the most spectacular glow that's only to be seen in the dawn sky has to be a morning comet. These are quite unpredictable and many amateurs spend a considerable amount of time trying to discover one.

Some comets appear only in a dawn sky depending on which hemisphere you live in. They might be seen in the evening then appear in the dawn elsewhere, or vice versa. These count as dawn objects if only seen where you live.

Many comets such as Hale Bopp in 1996–1997 were around for so long that they came into a rare category of those seen in the evening, night, and morning skies. See Figure 10.9 as it appeared in the late evening.

A true morning comet approaches the Sun in an orbit that only allows us to see it in our morning sky. These are very rare and good examples in recent years that don't need a properly guided telescope even more so.

Keep looking at news snips and be ready to go out and see the next one. It could be visible even tomorrow morning; many only show for a few days as they move in toward the Sun.

CHAPTER EIGHT

Daytime Glows

8.1 Introduction

Dayglow has come to mean a type of paint and clothing and is somewhat correct in our study as many of our lights do have a similar iridescent character. Many of these daytime glows have their counterparts at all times in the 24 hours.

For our purposes here we're only considering effects that are entirely due to the Sun and its light when the Sun's disk is completely above the horizon. At the poles this will be months' long during high summer, and not at all in midwinter.

A few of the daytime glows are so fascinating and varied that the whole of this book could be devoted to any one of these single issues. The halos and rainbows are cases in point, and because they can be seen at any time during the 24 hours they could have been included in an "Anytime" chapter. Daytime is when most people expect to see them.

The best time to look around and anticipate the wide variety of glows on offer in the next few daylight hours has to be this short period immediately after dawn's merging into a clear day. At dawn itself the air is steadier and fewer people are about to get in your way, or complain of "loitering."

Nearly all glows will depend on the weather conditions and level of atmospheric pollution. By knowing "your enemy" in this way it is possible to predict and be ready to savor the events.

The basic rules are extremely obvious.

- *Wet or rainy*: Always be ready to anticipate lightning or rainbows as these are now particularly well seen toward the Western horizon. Is it worthwhile moving to get some nasty man-made structure out of the picture? Or will that enhance the artistry?
- *Cloudy*: Be ready for any breaks and see some sunrays or crepuscular rays.
- *Foggy*: It is likely to be less spectacular, which makes this a good time to practice observing specters and similar shadow effects. Whenever there are bright artificial lights still on, setting up now or checking conditions for a night shot makes sense. In a particularly overcast day check whether or not dusk phenomena appear.
- *Blue skies*: Whenever there's some haze or high wispy clouds, then anything can happen. It usually does.

In other words, never give up and always look up. And always carry a camera with you.

You must always check what's going on above you. To a glow-watching beginner, the biggest surprise is their sheer size. Many daylight lights and glows can take over the whole sky.

We've had a good indicator of what's to come from the dawn colorings and must now start looking systematically for those only seen in daylight, although some do have their mirror or sister in full moonlight at night.

We start by looking at an early morning sky soon after sunrise.

8.2 Atmospheric Pollution

This is such an all-pervasive topic that we must let this take priority in any expectation of day glows. The main features of atmospheric pollution are so important that a separate chapter (Chapter 3) has been devoted to them (Figure 8.1).

It's now an unfortunate fact of our modern life that quite a number of sightings can only take place because of the "nasty stuff" in the air. The really sad thing that goes with it, as we've seen at dawn, is how many of the fainter or indistinct glows have been totally obliterated.

8.2.1 Mechanism

The atmospheric pollution can be entirely natural, like volcanic or desert dust, or entirely artificial, from forest burning and car effluent. In all cases it's something that everyone should be warned about because of the health hazard.

Always make use of this valuable information whenever it's announced and go out and see how this affects ordinary sunlight.

When there's a lot of this pollution has now, unfortunately, become the best time to get outside for the more spectacular views of sunrays and crepuscular rays.

Figure 8.1. *Eclipse, Arizona, 1994* When the air is clear of desert dust, miniscule glares around the sun allows me to catch an eclipse.

8.2.2 What to Observe

The atmospheric pollution that is to be paid most attention to is an entirely new one starting in the last century.

8.2.3 Contrails

To some these are a complete disaster as many a night's observing session has been blotted out as these trails cover faint objects. I'm not going to disagree with that opinion, as this has become a main contributing cause in the loss of the night sky. In the daylight this has become our multiedged weapon.

- The effects on climate remain to be resolved, but at ground level a heavily screened cloud cover due to spreading contrails can lull the public into a false sense of security. Contrary to common belief, such clouds do not diminish the risk of sunburn as ultra violet light passes right through.

Our immediate interest is in glows and how ultraviolet light interacts in any we see and makes it into a potentially fascinating research topic, another good reason to study contrails for the science that can come out of them.

8.2.3.1 Short Trails On a crisp clear day contrails will be short and dissipate within a few seconds. This is a brilliant day in all other respects. Do not expect to see glows. This is a much safer test of atmospheric clarity than the "Bishop's Ring" when pollution is known to be absent.

8.2.3.2 Thin Trails

These are the normal WWII style that go from horizon to horizon but do nothing spectacular other than gradually dissipate. If seen in an otherwise clear sky, again glows are unlikely, but you never know. . .because they often presage:

8.2.3.3 Expanding Trails

More often than not, these are the best harbinger of bad weather on its way. Glows of some type are extremely likely and dusk's prospects exciting. This prospect almost becomes a certainty if the expanding contrails merge into a complete haze with cirrus and other "wispy" cloud formations approaching from the West (Figures 3.8 and 9.4).

Before high-flying aircraft appeared on the scene you could have never seen single sun dogs appearing as if out of nowhere, nor the multitude of shadow and corona effects from those trails and aircraft themselves on these extra, but unnatural clouds.

A few windless days with strong sunlight and the effects of town-generated air pollution rapidly generate modern smog.

Put on your smog masks and venture out to add to the tally and prospects of seeing another group of glows.

8.3 Bishop's Ring

A genuine Bishop's Ring is extremely difficult to spot and is also the most potentially dangerous to your eyesight. The danger is not that obvious because the main use of the phenomenon is inverted logic, that is to say, seeing that it isn't present. That's why the dangers lurk because to be absolutely sure it's necessary to stare for a long time directly at the Sun.

As they say in all TV programs—"never try this one at home."

Modern video and digital cameras do the job just as well as a direct view with the naked eye.

It used to be said that to see whether or not the effect is there, "hold up your forefinger and get the tip to cover the Sun's disk." A much safer alternative is to make use of a lamppost or a similar thing with an end just small enough to just cover the solar disk, and then move so that it does that (Figure 8.2).

By far the safest way to check this glow, or lack of it, is to view a reflection in a window or puddle, preferably into dark glass. It will be obvious immediately whether or not there is a bright haze around the Sun (Figure 5.3).

I've included it here because it is a genuine effect, normally only seen during the day. Little did I realize that the picture I took in 1992, subsequently used for the Eclipse book a few years later, would be a perfect Bishop's Ring example. The orange arc around the set-Sun point is clearly obvious (Figure 8.3). See also Figures 7.3, 9.13, 9.14, and 9.15.

Figure 8.2. *Newcastle, 1996* Carry a camera all the time, underexpose a little, and a corona adds impact to a familiar street scene.

Figure 8.3. *Book cover* Only when researching this book a decade later did I realize the significance of the cover illustration used for "eclipses," the Bishop's Ring from the Pinatubo eclipse a year before.

8.3.1 Mechanism

On an otherwise cloudless day when the ring is a possibility, the glare around the Sun will be too bright to tolerate and that might be due to the following reasons.

8.3.1.1 The Real Thing To see the effect at all, the atmosphere must be completely clear from all the normal clouds and pollutants. Otherwise we see a corona or similar effect. Or the normal water clouds swamp the effect.

Bishop's Ring is derived from light scattered and refracted from extremely minute particles of volcanic debris in the upper atmosphere. All the larger particles have settled out long beforehand and these are too small to be affected by normal gravitational effects in the jetstream and similar winds operating in the stratosphere. Most of the stuff will be ash about a micrometer or smaller in size. Mixed in with this will be similarly sized aerosols of chemicals left over from the eruption.

The chemical composition is a research project in itself and the bulk is mainly sulfuric acid and similar sulfur (sulfate) compounds that lead to dramatic global cooling by reflecting sunlight back into space.

The particles are too few to dramatically affect the apparent sunlight reaching the ground but there is sufficient scattered light to create a faint white haze around the Sun. This haze disk in bad cases just after an eruption will spread out to well over a high percentage of the sky. In the course of time the disk will shrink and finally vanish as all the debris finally gravitates down to the troposphere and gets washed out in rain.

8.3.1.2 Other Causes Any minute particulate matter in the upper reaches of the troposphere and above will generate the effect. Fortunately these other sources are rare and mainly confined to regions close to ground level and rapidly wash out.

During high summer always check the weather forecast for hay fever warnings. Masses of pollen often reach surprising heights and density, more than enough to create a Bishop's Ring effect. It is worth noting that some pollens don't appear when expected and those from trees are extremely common in early spring, whilst spores from fungi predominate in the fall.

8.3.2 What to Observe

Around this haze and glare, but within the boundaries, will be a barely discernible colored ring of light. This is normally yellowish on the inside to a pale red to orange outside.

That's the real thing.

And why so difficult to spot as the eye must accommodate to the brightest glare imaginable and then struggle to sort out colors at the edge of vision.

My *strongest* advice is never to attempt this glow's sighting but to use the logic in reverse.

8.3.3 Sky Clarity Guide

In the twenty-first century the chance of finding a day without some atmospheric pollution is becoming very rare indeed. However, if you are fortunate enough to live in a clean environment, the quickest of checks for a Bishop's Ring becomes an extremely good sky clarity guide.

Try holding up your forefinger to block out the Sun and take the briefest of glimpses in your video screen with the lens set at a modest telephoto setting (NEVER attempt naked eye). You'll need to screen the lens with something whilst setting this up and a tall lamppost does make this chore a lot easier.

If the screen isn't immediately blanked or washed out with the video's safety cutout, then the chances are high that the sky is extremely clean and there's little or no pollution to scatter light. The sky right up to Sun's disk should show as bright blue with no whitening or glare.

When this happens you are extremely lucky indeed and many of the fainter nightglows should be fully visible later on.

Combine this quick method with a check on the rapid contrail dispersal and that makes this combination a powerful tool in assessing sky clarity. The snag is that most of the interesting glows become impossible under these clean skies.

8.3.4 Green Suns

Sometimes the aerosol contamination in the stratosphere is of the right type and size to scatter enough of the spectrum to show the Sun as a greenish tinge. Sometimes, and much rarer still, the scattering is enough to show a blue Sun.

This is the daytime equivalent of a genuine blue moon.

8.4 Sun dogs

Very much to be admired, these transient bright lights on either side of the Sun are by far the most common glow spotted by the public. They are a special case of the multiplicity of general halos caused by sunlight passing through ice crystals in the upper reaches of the troposphere and lower stratosphere.

They are most regularly spotted shortly after dawn when low on the horizon, or shortly before sunset. They are not rare and can appear at all times during the day; and at night when the Moon is up, they're called moon dogs.

Sightings at noon are rare simply because only few gaze up that far. No sun dogs are possible when the Sun is at more than 61° elevation.

The much more common name for these bright lights comes from their appearance as *Mock suns*. The scientific name is *Parhelia* when both are seen and *Parhelion* for a loner. Literally translated as "by (or with) the sun."

8.4.1 Mechanism

Parhelia always seem to make a surprise appearance when the skies are particularly thinly overcast. The sunlight at the time gives a feeling of being chilly and is definitely pale. The reason really does arise from those chilly conditions in the upper atmosphere; the clouds are a myriad of tiny ice crystals (Figure 11.1).

The most common ice crystal shape is the plate form. This is also most likely when water vapor freezes rapidly, as from an aircraft exhaust in a contrail. Single sun dogs are very regularly seen when the contrail sweeps into the right position which is 22° to either side of the Sun, and at precisely the same elevation (height) above the horizon.

Whenever upper atmosphere conditions are favorable, these plates stop tumbling in all directions and have a tendency to drift downwards like a sheet of paper. The effect is to see the majority of these crystals horizontally.

When enough stay that way, sunlight is able to pass through one crystal edge (face) and is refracted through the 120° opposite face to appear deviated through 50° in all. The net effect is to show up as a bright light 22° on either side of the Sun.

These sun dogs are always part of a total circle of light, the full 22° halo (Chapter 11), but that's rarely seen. There is always the possibility of solar pillars (Chapter 7), again extremely rarely seen.

8.4.2 Shape and Color

As normally seen, sun dogs show up as a short column of light on either side of the Sun and no more than that. The height is rarely more than a solar diameter high.

For a full-blown apparition, the height will gradually merge into the full circular halo, which means there are no actual limits on the height, just that the column becomes more arched above and below the Sun's elevation.

Under these conditions the width will broaden outward, invariably never inward toward the Sun, and this will rapidly taper off to a point, sometimes several solar diameters long. Needless to say, these extended appendages outward are often referred to as the "dogs' tails."

The majority of the indistinct sun dogs show little or no coloration. In really clear skies the inner band can be bright red, which quickly expands into the full rainbow colors before merging into a white to yellow tinge.

Examples are shown in Figures 8.4 and 8.5. (also see Figures 9.6 and 11.7).

8.4.3 What to Observe

Because sun dogs can appear at any time when there are wispy clouds about, it pays to carry a camera with you at all times. To take a picture, the trick is to shield the lens from the Sun by some convenient object in the foreground. A tree is best, although manmade objects can enhance the artistic effect. It is vital to ensure that the camera remains horizontal because the sun dog to Sun line is precisely that. Failure to do that simple thing shows up as a sloppy work!

- Sun dogs created by contrails make nice pictures and only add to the science when trail dissipation rates are logged.

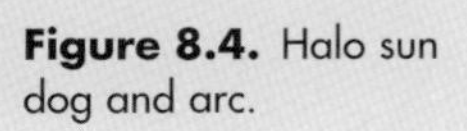

Figure 8.4. Halo sun dog and arc.

- The angular distance isn't exactly 22° at all times. Check to see by how much this increases as the Sun's altitude rises.
- Sun dogs seen through sea and other mist can persist right down to the horizon. The bright colors merge into the general red sunrise and sunset. It's rare for both sides to survive under these conditions as clouds of a more normal type intervene.

The final piece of observation to carry out is extremely simple.

Concentrate on recording these mock suns every time they're spotted, and log the sky conditions and subsequent weather. Keep notes on the weather for the next few days.

- Whenever a single sun dog is seen, or a pair that last but a few minutes, they are often nothing special; a pretty sight, nothing more.
- Whenever a "matched pair" of sun dogs is seen at a modest altitude, check to see how long they remain visible. Those that come and go rapidly as above

Figure 8.5. Very bright sun dog, Bolivia, 1994.

are of no more interest than any other transient glow. Those that remain bright and cheerful for a considerable period, certainly more than half an hour, then persisting for hours are particularly valuable in their use as a weather predictor.

Persistent sun dogs with what are often called "tails" pointing outward and associated with a quite definite halo are created by ice crystals usually associated with a weather front. That's usually a nasty one lasting a day or two.

The best example of this I've seen was in Sumatra, right on the Equator, where ice crystals in the clouds shouldn't happen. The heavens opened for some time afterwards and the whole of Asia was, it seemed, inundated with floods, from the Himalayas downwards. That was in August (summer there) of 1998 (Figure 8.6).

Persistent and really bright sun dogs, seen more or less on their own, that is to say no trace of other arcs or halo, are created by a special concentration of another right type of ice crystals that precede an anticyclone. All the memorable ones I've seen have been followed by days, often a week or more of extremely settled, bright sunny days; winter or summer that seems to happen regularly.

Slap on the sunblock, else you'll get that nasty sunburn glow. . . .

Figure 8.6. Ice clouds sun dog at equator, Sumatra.

8.5 Corona

Coronas are not halos, although they look very similar at first glance. Their origin, however, does lie in a similar source, that is to say from light refracted round water droplets, ice crystals and general atmospheric pollution. See Chapter 11 on halos for an explanation of a few of those many possible permutations.

The Bishop's Ring is a true corona seen in its simplest form. In the case of the Bishop's Ring we have nothing between us obscuring the Sun other than a very fine aerosol at a very much higher altitude.

8.5.1 Mechanism

A true corona never arises from light transmitted through something, or from reflections. That statement inevitably leads to some confusion because those two factors must always happen at the same time.

Light from those two refraction and reflection sources is the origin of most of the central bright light patch at the center of a true corona. The corona proper only contributes a proportion of that central lighting and is never seen close to the center for that reason. Refracted and reflected light simply blots out any chance of seeing anything that close in.

The corona occurs whenever the air contains a lot of particles small enough to diffract sunlight. These are around a micrometer (micron) size and can be anything. The name for particles in this size range is aerosol, which is actually a lot smaller than from most medical dispensers.

- The normal and completely natural particles are water vapor condensing as exceedingly fine drops, but much commonly as ice crystals formed by a sudden cooling. After that source, pollution and volcanic dust feature highly although some authorities think meteoric dust from space might be a reason sometimes.
- The next requirement for a corona to be seen is an extremely uniform particle size. Once the particles vary in size by more than a few percent, the different-sized coronas they produce and their individual color bands all overlap to make a normal bright splurge of light.

That's all you see from normal clouds containing a whole hodgepodge of particles.

- Provided the particles are uniform in size and distribution, light refracted round each particle will add to the general effect.

The delightful effect is one of a distinct dartboard of colored rings centered on the Sun, around a central white patch.

- The colors seen do not conform to the rainbow sequence and are counterintuitive.

- To add extra confusion, the pattern is repeated more than once to finally peter out into the background sky. The colors aren't due to rainbow reflections but refraction, and each particle contributes its own ring. What you see is a summation of all these colors overlapping and that can never be a pure color, rather a consistent pattern of reproducible dimensions.

  The corona's overall size is entirely dependent on two factors:

1. First, the particle size and uniformity of that size range.
2. Second, the distance from you to the cloud containing them.

From this it follows that a deep cloud can't give a corona as there are unlimited possible sizes from different distances, which then overlap to cancel out to a uniform whiteness. That's why I say it's what you see in any normal cloud.

Aircraft contrails spreading out make super possibilities for a corona because of the incredibly uniform exhaust on a cruising run. It is somewhat ironic that this atmospheric interloper has an attractive plus side.

8.5.2 What to Observe

As in all occasions, when viewing into the Sun directly, safety is paramount. Read the notes on Bishop's Ring.

Always ensure some tall or substantial object blots out the Sun.

The color sequence gives this glow away. The first color out and bordering the whiteness is yellow, merging immediately into an orange, then red, violet, blue, and green to be repeated outward.

- Do you see anything like this sequence? If so, how many repeated patterns?
- If you photograph it, the repeats should be about twice the diameter of the next one in. Whilst not easy to identify where one color starts and ends, this crude test will confirm that you have a corona.
- What you see might not always conform to the colors recorded by film or digital as the three sensing devices won't be the same, nor have the same relative sensitivity to each medium, nor to your eyesight.
- Because most sightings occur through thin cloud cover, the dramatic size changes as each cloud passes, making this sighting much more dynamic than most daytime glows.
- Quite often the corona is a lot more obvious when wearing polarizing sunglasses.
- Shatterproof car windscreens sometimes show this effect when wearing polarizing sunglasses.
- A final, almost foolproof way of seeing a corona is extremely easy to set up. A fine mesh net curtain made from some artificial fiber with sheen does admirably. Simply get behind one in a window, or erect one, and view the Sun on a crystal clear day through it. The first time I saw this effect was in a hotel whilst waiting for a flight.

I know this is a complete fake, but it is the best way I know of seeing a corona without all the hassle of setting up in the open air, with madding crowds to interfere. Note the corona size and colors and how these vary with your distance away from the curtain (simulated clouds) and how the mesh size affects them.

Figure 8.7. *Eclipse in Chile, 1994* An example of a solar corona surrounding a solar corona during a total solar eclipse. That this is a genuine effect is seen from the corona around Venus, *top right.*

- Although nothing to do with the daytime, a corona is often best seen at night as a colored 'halo' round the Moon, when fine cirrus passes by. See also Chapter 4 on the weather, and Chapter 7 on "Red Sky" (Figures 8.2, 8.7, and 8.8).

Figure 8.8. *Lunar corona, 2006* The coronal coloration is much better seen at night, around the moon.

Whenever the air is particularly clear and there are thin and wispy clouds about, get ready to "crown" your list of glows with this one.

8.6 Glory

A glory is nothing more than a special case of a Brocken (Chapter 7) and is the terrestrial daytime equivalent of the Gegenschein seen in the night sky (Chapter 10).

8.6.1 Mechanism

The basic requirements are the same for those other much rarer glows.

- Unobstructed sunlight shining past you onto a reflective screen of particles. Your shadow appears to be immersed in a bright circle of light merging into and surrounded by bright rings of colored light.
- The reflective screen can be anything and is normally a white cloud of water droplets, ice crystals in the same clouds, or fog and mist, but any uniform ground can work almost as well.

It follows from these basic requirements that the best place to see something is a high place looking down into a cloud or mist. Mountaintops were the only viewpoints until last century, now everybody should expect to see one sometime during an aircraft journey.

Sunlight around the shadow is reflected back from the clouds and the apparent size varies from huge, when close by, to invisible at heights much above a half mile/kilometer. The difference from a Brocken's specter is this lack of an obvious shadow linking you with the ring structure.

Other than that briefest of introductions, the true mechanism remains a matter of some considerable debate. The coloring can be explained as light refraction around you, your size acting as though it were a tiny obstruction at this scale of things. Something like that must be the case when a glory is seen on brown earth or green grass.

The alternative and most likely explanation for the most frequently seen glories must lie with light reflected back from cloud particles as occurs in the Brocken. Then the droplet size and uniformity must come into play in deciding the brightness and intensity of the central region within the colored bands.

A combination of all these modes hedges all bets to explain those I have seen.

Sometimes the glory has its normal bright central region, and at other times a dark center. The size of the central shadow or spot also changes to an amazing extent and can actually spread out beyond the colored rings. That's why you can carry out some original scientific research by keeping your eyes open.

8.6.2 What to Observe

Next time when you're in a plane high above the clouds always check to see if a glory is on view.

By far the best viewpoint is a high-wing light aircraft. During my regular trips in one, seeing a multitude of glories is an extremely pleasant way to pass the time, and a good excuse for using lots of film and video stock.

Whenever the sky is clear above the plane and it's close to clouds, the conditions are near perfect, as they were one day. We, the passengers, were treated to a brilliant display as we passed through cloud cover and saw glories all the way down through the many clouds to ground level (4,000 ft) and onto the grassy ground. This time the plane's shadow exactly matched the ring size (Figure 8.9).

- The Sun's altitude doesn't seem to matter, as on that day close to local noon, as long as it's shining bright and clear.
- The ultimate "show-off" is possible in a light aircraft. When passing through the top of light cloud cover into the Sun, a genuine corona in front may well be matched by a glory behind. The practical advantages of a light aircraft are then apparent when traveling across the Sun's path as it is then possible to see both phenomena by looking at either side.
- When skipping along the top of clouds, glories will be huge and appear to touch the plane's shadow, converting this to the same optical delusion as

Figure 8.9. *Triple glory* More than a single ring is very rarely seen; an example of catching the opportunity when you can, even through a dirty window. This is a genuine picture, not one manipulated for "effect."

a Brocken, confirming the same mechanism. Colors will be vivid and in a constantly changing ring size and distance.

- The great mystery surrounding glories becomes apparent in a long flight over different cloud types. Sometimes the glory is much larger than the plane's shadow, at other times seen well within its confines.

A possible explanation has to be in water droplet size, with raindrops ready to fall mimicking a rainbow. However, that doesn't explain the same variation over dry land.

- Sometimes when a glory is seen in a real mist, there might be a much fainter fogbow outside the main glory. If that is ever seen, then the rainbow explanation has some substance in these rare sightings. Unlike the ever-changing glory size, a fogbow will always extend over the same rainbow's 42°. To experience this an observer or plane must be well away from the fogbank.

At the extreme height of modern jets it is still possible to pick up glories on the ground, which indicates that diffraction effects of some kind must apply here. Glories are extremely attractive over ice and snow that act as a marvelous reflector.

On these same ice and snowscapes, it should be possible to pick up the contrail shadow. Deciding what color this looks like is highly dependent on window clarity, but is it a glory? I doubt it as the contrail isn't at the antisolar point.

Keep records every time a glory is spotted, and the backdrop. It will help unravel exactly what is going on.

8.7 Iridescence

Anyone who's seen an iridescent cloud never forgets it.

The first surprise one gets is to see a cloud that's not white, rather a whole raft of "shiny" colors. Hence the name "iridescent", a description often referred to as a metallic finish.

The second feature is the intensity. With some sightings the glare is so intense that your instinctive "blink reaction" kicks in immediately and you to want to look away as you blink. That's extremely good advice, because, as with all glows seen close to the Sun, eye safety is paramount and this is an occasion when double precautions are desirable. The first is to stand in a place where the Sun's disk is fully hidden behind something. The second prudent step is to wear the best sunglasses you have or get a low shade number welding glass.

Another useful tip, but not easy to set up on the spur of the moment, is to view the sky by reflection in a pool of still water. A sheet of clear or dark glass is an easier thing to carry around or have in hand than a bucket of water. Don't use a mirror.

8.7.1 Mechanism

The effect is always seen close to the Sun, hence the mechanism must be akin to the corona.

- Those seen within a few degrees closest to the Sun are invariably white, and those further away exhibit a very wide range of colors, strengthening the view that the corona explanation has some merit.
- The main difference lies in that iridescent clouds can be seen well away from the Sun and are easily mistaken for sun dogs. As these colors can appear anywhere, not restricted to precisely horizontal and at 22° on either side, they are also clearly not the same glow as a halo or our familiar sun dogs.
- These clouds can be seen at anytime of day and have their counterparts in the night sky.
- However, they do seem to be most common in a northern fall, suggesting that the changes to colder weather play some part in the cloud formation and the particles' size, shape, and abundance within them.

On most occasions, not all the clouds are affected and the brightest coloring is seen as patches, bands, or at cloud borders.

On balance, iridescent clouds seem to arise from the same mechanism as a corona, with the particles responsible being particularly abundant and/or uniform in character.

8.7.2 What to Observe

- Polaroid sunglasses make a very good tool to confirm these glows. The bulk of the background glare is cut out, whilst the iridescent cloud color is not modified and stands out more obviously. If the colors change markedly or vanish, then some other factor is operating here. Iridescent clouds are not polarized.
- The coloration seen can range widely with a blue cloud in a blue sky being the most unexpected. Purples are more familiar from sunsets and impending storms, and reds likewise. The characteristic difference is their localized nature and intensity.
- Heed the warnings not to gaze at the white clouds close to the Sun. Genuine Iridescent clouds more than 10° away will still be too bright for comfort in most cases. The test for these is to look away quickly whilst closing your eyes. If after-images of the same hue are seen, this becomes another good indicator. Really bright ones might even give the complementary colored after-image.
- Iridescence is in no way restricted to a specific cloud type. All of them at different altitudes can exhibit this glow at some time or other, possibly the same mechanism applies at night.

Figure 8.10.
Iridescent clouds. 2006
The most unpromising solar shield will serve to view this effect in safety.

- That the phenomenon is related to a uniform particle size or shape is clearly indicated by the frequency these colored effects are seen in contrails. Figures 8.10, 8.11, and 8.12 are more normal daytime examples, seen in August 2006.

The most beautiful iridescent clouds are much more comfortably seen at night around the Moon. Moonlight is both pleasant and safe. Sorting out iridescent clouds from all the other moon glows is a real challenge.

Figure 8.11.
Iridescent clouds, 2006
Catch these brightly colored lights in the sky whilst you can, as earlier in the day under a roof.

Figure 8.12.
Iridescence and gannets
This is much more likely an example of iridescence and, due to heavy cloud cover, associated effects, such as "mother of pearl," normally seen in different times of the day.

8.8 Mirage

Mention the word "mirage", and the immediate reaction is to think of a desert. These certainly feature as an inevitable function of the regular solar heating over large tracts of land. The appearance is extremely familiar to a shimmering lake between you and the distant horizon.

8.8.1 Mechanism

8.8.1.1 Inferior Mirages Hot-air mirages rarely reach enormous proportions. The hot air forms a relatively shallow layer close to the ground which then disperses and cools as it rises into the much thicker main atmosphere above. The interface is in a constant state of turbulence, which aids the illusion of a choppy sea surface. The interface acts very much as a mirror with light reflected back from the much denser air above. To add to the confusion, this mirage is described as an "Inferior" mirage, from the reflection back from above.

What you see is an image of the sky itself as reflected. That begs the question, which way up is the sky? The number of hot and cold layers and the number of times reflected are never constant.

- The inverted mirror effect explains why vehicles in the distance on a flat road look as though they're floating on air. In extreme cases their wheels stretch out to make it look as though the cars are running on stilts.
- This is also the explanation why the setting Sun can be "late." The real thing has long gone but the mirage allows objects well over the horizon to appear as though above it. This is also the normal explanation of a green flash (Chapters 7 and 9).
- It also explains many glow sightings. Snow-capped mountains hundreds of kilometers away can pop into view this way, as can clouds and similar

atmospheric features. At that distance away the lighting conditions will be totally different from that seen locally and add to the bizarre sight.

These strange sightings have occurred all through history and are a very likely explanation of many of the "visions" seen in the past, one idea put forward for Biblical revelations.

When these hot-air layers lie above a colder region at ground level an alternative situation occurs known as temperature inversion. Another mirage type then becomes possible.

8.8.1.2 Superior Mirages Mirages do not have to be in a desert country.

- They occur just as frequently wherever there's a large patch of cold air at ground level. Light appears to bounce off the upper and warmer surface, hence the name. Objects far away can then appear well above the horizon as if by magic.

In a temperate climate "Fata Morgana" has gone into our language and heritage. This is named after Morgan le Fay, King Arthur's half sister of legend who was said to be able to fly and change shape. A mirage created by layers of warm over cold air. As I've just described these layers bend light from sights in the far distance, making them appear distorted into what looks like mountains and castles.

Although a legend, this is clearly based on personal experience by someone seeing something similar. For more explanations on UFOs, see Chapter 14.

The Father Christmas legends with "flying reindeer" clearly indicate that mirages aren't restricted to hot climes as the same fantastic castles and spires magically appear over the horizon, often said to tower over his grotto. As long as there are marked temperature differences in separated air layers, with at least one warm on top, this glow is always a possibility. Katabatic winds tearing down from mountain ranges in Scandinavia and the poles fit this bill perfectly. Such an event changed history.

John Ross set out in 1818 to find the Northwest Passage but turned back when he found the way blocked by mountains. A second attempt by others a year later found absolutely nothing in the way. This mirage was such a familiar event locally that the Inuit even had a word for it – "Poojok", which translates into mist. This makes a great deal of sense because similar sightings in Europe have been reported after a hot day when looking out over the cold sea with a lifting mist.

Later in the century in 1897 similar sightings in Alaska of a city in the air were reported in reputable British Journals. The thinking at the time was of Montreal or Toronto, almost as far away from Alaska as London from those two cities!

- This type of mirage does have a name, Novaya Zemlya mirage, derived from the first scientific description. In that case the light from hundreds of miles/kilometers away was bent round the surface of the earth. It can appear to come from somewhere up to 10° below the actual horizon.

The classic example of this occurred at night. That was the sighting in New Zealand of Venus. Over a period of more than 10 days in December 1978, the planet was seen and filmed from that far below the horizon. The planet wasn't due to rise for the best part of an hour later.

8.8.1.3 Multiple Mirages Life is never simple and that applies to mirages.

- Whenever the conditions are favorable the cold and warm layers will remain in place whilst new ones are formed above or below to create a "wedding cake" effect. The possibilities are endless, as are the number and types of mirages seen.
- The most familiar manifestation of this is seen at sunset and sunrise. The Sun's image breaks up into layers of quite dissimilar size and coloring, quite often with nothing at all seen between some or all of these layers.
- These missing bands are referred to as a "Blind Strip" and can amount to a substantial proportion of the Sun's disk. The effect is likened to the Sun jumping or speeding up.
- These multiple mirages often lead to a whole series of green flashes, often seen underneath as well as above, as each layer pass through the boundary layers in the atmosphere.

The moral is that a mirage can be seen anytime and any place where there's a large and stable mass of air about. Visions and sightings of unusual as well as familiar objects floating in the air do have an origin in a well-established meteorological function.

8.8.2 What to Observe

Always expect to see strange lights in the sky during the day from places well over the horizon, but in their darkness, as urban light pollution expands. The time to be particularly vigilant is whenever the atmosphere is exceptionally clear and free from mist and haze, and distant objects appear to be right up close.

Some other glows to seek are as follows:

- Localized "hot spots" can give an inverted mirage where distant scenes appear as though below ground. The converse is also possible with nearby scenes upside down seen on walls.
- The lower you are the more dramatic the mirage appears when it's only the thin layer above the ground involved. As your eye gets down to ground level, the intervening "sea" will appear to rise up and "drown" you.

What you see in the sky immediately above that layer can and does contain all sorts of ghostly glows.

- The Sun by day and the brighter night objects can also be seen reflected back from the local hot spots. Always expect this over hot chimneys and exhausts.

Figure 8.13. *Mirage* Air layers at different temperature lead to this classic "vase" effect. The multiple image and halo effect is due to photography through double glazing, necessary in Iceland's cold climate.

- Conversely, always expect the Ross experience when on board ship in cold climates. This is the classic temperature inversion with a rapidly increasing temperature above. Distant scenes rise up and are seen from a very long distance.

Expect these mirages to change shape and position extremely rapidly, although in colder climes with more static air conditions this is less likely than in warmer ones (Figure 8.13).

A good time to experiment with mirage effects occurs whenever a car stops. The hot air over the engine or roof is a perfect layer bounded above by much colder regions.

Maybe this is the origin of the "Cat on a hot tin roof" expression?

8.9 Fogbows

There is absolutely nothing special about a fogbow. They are always seen whenever the Sun is behind you and not so high in the sky as to show a Brocken or glory in low mist banks below head height. Regard these as the more general case.

8.9.1 Mechanism

The mechanism of formation is exactly the same as in a rainbow with the minute droplets sending light back to you

- The essential difference this time lies in the droplet size being so small that diffraction effects predominate, rather than adding a tiny amount to simple reflections seen in a supernumerary rainbow (Figure 12.4).
- The net effect is that the diffraction colors tend to cancel out and destroy the rainbow proper. What you see is a white bow with some coloration if you're lucky (Figure 8.14).

Figure 8.14. *Fogbow, Egypt eclipse, 2006* Unlike a rainbow, a fogbow is essentially white and the end touches the ground immediately in front of you.

- The spectral sequence mimics a rainbow with the inner part bluish to violet and the outer more orange than red.
- There will be secondary bows and arcs, normally seen on the inside with a reversed color sequence.
- Another characteristic to be expected is seen in practice, in that the diameter varies considerably from a rainbow's 42°. Diameters up to twice this have been reported down to almost half that.
- As with all these phenomena associated with reflections from small objects, the center of the circle (we only see that portion above the horizon) is the point on the opposite part of the sky from the Sun. Camera pictures taken at the same time as naked-eye views readily confirm this similarity through all of the glows, the viewpoint centered on the antisolar point.
- If the fogbank is particularly dense and you are right up close, or even in it, the apparent diameter expands very quickly to become a full circle in extreme cases. This is precisely what happens when a glory is seen as an aircraft goes into a cloud. That confirms a similarity in mechanism.

This special case of a complete ring is given a special name, "Ulloa's Ring". To see this well the fogbank ought to be in front of a relatively darker background.

The really interesting feature arises when the water droplets freeze into ice. The crystal shape is highly unlikely to be spherical, and then surface reflections predominate. That is a partial explanation why some other optical features appear at the same time and add to an overall muting effect as the reflected colors are now stronger and cancel out the diffracted ones.

8.9.2 What to Observe

The main visual differences from a normal rainbow lie in the muted colors and low intensity. The reason for this is obvious in the filtering effect of the mist between you and the main mist or fogbank.

Figure 8.15. *Fog and sun dog* Instead of generating a fogbow (because sun is in wrong position) this sun dog appeared.

- Fogbows are always best seen when sharply defined fogbanks swirl in from somewhere. Sea mists are well worth watching, something happening today as I check this chapter, but that only showed a sun dog (Figure 8.15).
- Seeing more than the primary bow is extremely rare and always worth recording.
- Whether or not low cloud can be regarded as a fog is a moot point but high places often attract one or both.
- Whenever a temperature inversion occurs fog will often form at the boundary. When this happens fog may spread into both of the air layers.

The opportunity then arises to see a rainbow with a distinct join in it at the boundary. The brightness and intensity in each part is markedly different, and always strikingly unexpected. The best example I've seen was over Antarctica before the 2003 solar eclipse (Figure 8.16).

8.10 Rare Daytime Glows

Lightning is not a rare daytime experience. This is best regarded as a glow that can happen at any time (Chapter 13).

When conditions are extremely stable you might be lucky and see the Heiligenschein (Chapter 7) well into the day and follow it right round to sunset.

Sunrays during the daytime are so common that they hardly ever raise comment. Sightings around noon tend to be missed from our natural reluctance to gaze up that high. These are best seen when the humidity is high immediately preceding or following rain.

Whenever the dust level is high expect to see these sunrays and crepuscular rays at any time. Volcanic dust has been considered under Bishop's Ring.

Figure 8.16. *Bows over Antarctica* A rare arc was seen at 35,000 ft over Antarctica immediately prior to a total solar eclipse. The change from clear air ice crystals overlying thin cloud breaks up the complete arc formed by sunlight shining on them.

One of the more intriguing daytime glows is also amongst the most worrying, as it's an extremely good indicator of imminent and the nastiest conditions imaginable.

8.10.1 Green Skies

This forewarning is to be expected wherever tornados are common, as in the United States, but can happen anywhere there's a severe storm.

The classic events in the days April 3–4 in 1974 were quite remarkable when no less than 148 tornados in that short space of time careered across the mid states in the United States. The death toll amounted to 330, at least 5,484 severely injured and an unknowable cost of structural damage. Nor is this experience confined to that geography as green skies preceding severe storms are increasingly common in Britain in its own "Tornado Alleys" around Birmingham and Selsey.

8.10.2 Mechanism

Quite why the sky turns a bright green, almost emerald, instead of its usual blue remains a mystery.

The answer has to lie in the huge amount of water vapor carried in these very warm storm clouds together with astounding quantities of large water droplets and huge hailstones. Somehow sunlight is absorbed and scattered to leave a predominantly green hue.

That the coloring is real and not an optical delusion or contrast effect has been amply confirmed by the University of Oklahoma, who were able to get the first scientific recordings in 1995. Green was the correct description for the sky coloring.

8.10.3 What to Observe

Whenever the weather forecast is a severe storm warning and the sky turns green, make a quick observation, then seek secure cover....

Expect tornados, damaging hailstones, and always thunder accompanying lightning damage.

This is one glow that's never going to get me green with envy.

CHAPTER NINE

Dusk Glows

9.1 Introduction to Dusk Glows

After a hard day's work has stopped you from going out to see the daytime glows, what better way to relax than to settle down with a favored beverage and watch a gorgeous sunset? Why is it that we all tend to think of the sunsets as the only time to view our glows? Maybe because of this freedom from the day's worries and stresses as we get ready for the night watch.

The transition from day to night follows the same pattern as dawn in reverse but there are some important variations. Top of the list is the amount of water vapor in the air.

During the course of a normal day this will rise to a peak around mid-afternoon and fall back steadily to dusk. This is quite definitely not the same as relative humidity (RH), where the same amount of water vapor corresponds with a lower RH figure at higher temperatures. Dusk temperatures tend to be much higher than at dawn, a quite significant difference that's seen in a considerably lowered RH.

When the sky is free from weather fronts this goes a very long way to explaining the more dramatic effects in the failing light. Light cloud cover can form, then disperse just as rapidly leaving the atmosphere appearing quite clear but full of that invisible chemical-water.

It is only under thunder conditions when even the hot air can no longer hold more water vapor that the more spectacular thunderstorms and lightning displays take place. These are quite rare at dawn. This topic is taken under glows seen at any time (Chapter 13), but refer to Green Sky (Chapter 8) for a special case.

This extra water vapor has little effect on some of the glows, and we can say that crepuscular rays, earth shadow, and all the dusty phenomena are merely enhanced with an extra reddening. Volcanic-dust sunsets always impress this way.

That leaves two sunset glows that are much improved and exhibiting a wider range of features.

These are solar pillars and the green flash.

Then we always expect there to be a "Red sky at Night, Shepherd's Delight."

9.2 Crepuscular Rays

Without any "shadow" of doubt the finest displays of these dark bands and associated sunrays are hugely enhanced as evening falls.

9.2.1 Mechanism

The conditions become perfect when the air is about to form cloud but hasn't quite made it.

Whenever there are small individual clouds about and the Sun is about to set, you're really in business. Don't waste any more time, it's time to be out.

9.2.2 What to Observe

- Listen to weather and similar forecasts for any special conditions such as dust, as described in Chapter 7, which enhance the prospects considerably.
- As with all the following glows seen at dusk, getting ready is so much easier. Set up your equipment and choose the best site in daylight with no fear of tripping over something vital.
- With these bands and sunrays occurring under lighting that we're familiar with, no special precautions are needed.

That is to say, don't look directly at the Sun when it's more than a Sun's width above the horizon, but only when the intervening haze or mist diminishes the disk's brightness. Never look through a normal camera lens or telescope without consulting an Astronomy Society or other experts on the correct or best filters to fit to the optical aids.

This is a vital precaution, as the Sun will almost certainly come out from behind the cloud cover sometime. This is an essential action in polar or tropic regions, where the atmosphere is, otherwise, exceptionally clean.

- Taking still pictures never ceases to amaze with the sheer beauty possible. Always try to put some foreground interest in the picture but do remember to override the autofocus! It's the rays you're after not the tuft of grass (Figures 9.1 and 9.2).

Figure 9.1.
Crepuscular ray, Kenya, 1980 Shadow cast by Kilimanjaro as the sun sets into a dusty atmosphere.

- Video really comes into its own for this glow. Alternatively try a timed sequence from a fixed tripod. The rays appear to wave at you, sometimes even a "Mexican Wave."
- Never give up. Some crepuscular rays can be seen up to half an hour after sunset around the equinoxes in temperate climes when the Sun sets close to due West (Figure 9.3).

Our next dusk glow might also appear under similar atmospheric conditions to make the setting up and preparation doubly worthwhile.

Figure 9.2.
Anti-Crepuscular ray, Kenya, 1980 The other side of the sky view showing the pronounced parallax effect at the antisolar point.

Figure 9.3. *Multiple rays, Alderney, 2006* How sunrays and crepuscular rays are formed by individual clouds formation as moisture condenses at sunset.

9.3 Solar Pillars

The huge advantage of an evening watch for solar pillars is in seeing or in knowing the likely weather pattern. This can be in two ways, of course, and unexpected storm clouds appear as if out of nowhere, or the promising cirrus "burn-off."

9.3.1 Mechanism

In the Northern Hemisphere with its predominately SW wind flow, the appearance of wispy cirrus and the like coming out of the setting Sun is always a good indication of a good display. But there is a huge difference from the morning sky.

- In this case we don't have a Sun just at or below the horizon shining through a curtain of cloud. The cloud isn't moving toward the morning Sun, which is in a clear or nearly clear sky over the horizon. No, the setting Sun is shining through this approaching band of cloud before the light strikes the clouds we see.
- For this reason the sunset solar pillars tend to be shorter and much more muted than the morning affair.
- For really good displays the conditions must mimic the morning's and that means a band of cloud clearing the evening horizon and holding there long enough so that the Sun sets into a clear sky beyond. Timing is everything here, and normally known a few hours in advance with modern satellite weather forecasts. The higher the cloud the better as that greatly increases the "curtain effect," and with it the chance of the clouds containing the right type of ice crystal due to the colder air up there.

Figure 9.4. *Solar pillar and contrails, United Kingdom 1977* Sunsets tend to give redder backdrops.

As in the morning, we can also get the display from anticyclonic weather patterns when the air is holding enough daytime moisture to form wispy clouds as the Sun sets down to the horizon (Figure 9.4).

9.3.2 What to Observe

With the higher probability of a reddening sky from any of the reasons above and the larger amount of blue light scattering with all that extra stuff in the way, the contrast will be much lower. That's the muting effect.

- The great trick with an evening solar pillar is to take a picture with some foreground interest. I have the advantage of a boat-filled harbor with a large number of sea birds, but aircraft do a good job as a substitute.
- With sunspot maximum approaching as the first decade of the twenty-first century ends, try to capture sunspots.

Try your hand at all these variants of a hackneyed sunset and test the Camera Club Judge's knowledge. Most seem to think the effect is from a montage in Photoshop.

9.4 Sun dogs

Sun dogs and Parhelia are seen throughout the day (Chapter 8). What's so special about those at sunset?

9.4.1 Mechanism

Nothing at all except their positioning on the horizon, the frequency of reports, and their surprising brightness. Reports of two or more setting suns are quite common.

The atmospheric conditions for seeing mock suns at sunset are exactly the same as at any other time of day. You still need a concentration of the right-shaped ice crystals in the sky.

- The main difference at sunset is another simple piece of geometry as you're now looking through, that's along, a horizontal layer of atmosphere instead of straight through, that's across. The concentration of whatever is up there has to be higher.
- Provided some crystals are present that are too few for a normal daytime show, the scene is set for a possible good display.

Ideal conditions apply whenever the sky is "undecided" whether or not to put on a solar pillar or green flash. Those might happen, too, of course. As long as the horizon has a slight curtain of wispy stuff, we're set for a possible mock Sun.

9.4.2 What to Observe

These horizon mock suns at dusk rarely appear in matched pairs. One or other is muted or missing altogether. Hence the press reported, "Two suns were setting together." Strands of straggly cirrus and mackerel clouds coming straight at you, but a very long way away on either side of the Sun rarely have the same density as a single one seen directly above. The smaller but closer cloud bank will cover a much wider portion of sky.

- A good display is always on the cards when bright Parhelia appear earlier in the day.
- Plot their progress down and see whether or not they switch sides as the cloudbanks travel across the sky.
- When low down like this the glows might only show for a few minutes and never be seen again. Be prepared.
- As the glow approaches the horizon, the bright colors fade very noticeably, finally turning into a bright orange. This orange can turn a livid red just as we see in a normal sunset.
- Placed as they are like this on the horizon proper, it's now easy to see that the real shape is indeed a rectangle to square. That makes the sight an easy identifier to which one is the real Sun.
- In really nice cases it should be possible to see the "puppy dogs tails."

Figure 9.5. *Halo and sun dog, 1999*
Compare this picture with Figure 11.7. Underexposure helps bring out the coloring in this sighting of a sun dog's tail.

When these sun dogs set like this, it confirms they really do occur precisely at the same apparent altitude of the Sun. In a normal sky, perspective and vertigo distort our perception (Figure 9.5).

9.5 Green Flash

Often regarded as the "gem" of glow watching, the GREEN FLASH at this time of day deserves its capital letters here. Many readers will never see it from home because the conditions are so critical, not least an unobstructed horizon, very difficult in an urban environment.

9.5.1 Mechanism

I've covered the salient points in Chapter 7 on dawn glows, but the really critical consideration at either time of day is the need for the atmosphere to set up inversion layers for a mirage (Chapter 8). The mechanism by which the sunlight is bent around the horizon by refraction depends on one or more inversion layers.

- The dawn conditions are actually much more predictable, although rarely seen at that early hour. Astronomers in the Canary Islands often get up early because they can do just that.
- The other critical requirement is a really nice and clean atmosphere; normally the opposite conditions from rays and pillars.
- Whenever the air is exceptionally clear, get ready to think in terms of a green flash. Any horizon looking out over a relatively cool area, such as the sea, after a hot baking Sun is a good predictor.
- Next, steady wind conditions assist a lot. Warm winds blowing out from hot land over a cool horizon might help.
- The higher you are the better. A skyscraper in the fall in San Francisco looking out over the Pacific is a good substitute for the Rocky Mountains.

9.5.2 What to Observe

Because the sight is so much better with optical aid, the safety warnings must be repeated, yet again, with the added dangers from the clear air.

- Any standard camera will capture a good display but for best effect a focal length equivalent to 300 mm on a 35-mm camera is a basic starter. Anything boasting a lens with more than 700-mm focal length is so much better and my favored setup uses 1500 mm which half fills the frame. A tripod is absolutely essential.
- Most good quality video camera lenses will zoom out to a surprisingly useful image size. Use this lens zoom function, definitely never the image zoom.

9.5.3 Beware

Zoom lenses can generate all types of internal reflections when looking straight at a bright object, like the setting Sun. The more off-center you have that, the worse and more likely that optical defect(s) will show up.

The same warning has to apply whenever you fit a filter to the main lens. Again, multiple ghost images from this extra bit of glass in the optical system is guaranteed to set up even more ghosts of ghosts. Any grime or dust on the filter might also be locked onto by the autofocus.

Finally, do not, that is to say *never*, try to photograph or view the green flash through a window, particularly double glazing unless, of course, you want to fake it. See Figure 8.12 for an example.

You have already been warned. Many a picture I thought was a green flash turned out to be reflected (fainter) image from any of these common mistakes.

- Handheld binoculars serve very well but are never recommended with more than a fraction of the Sun's disk on view due to the inherent dangers. That warning applies in spades with any form of telescope.
- Whenever the Sun exhibits weird shapes and bands, then multiple green flashes must always be a possibility.
- Always expect to see the "negative" green flash with the coloring on lower segments instead of, or in addition to, the upper bits. The mechanism for superior and inferior mirage (Chapter 8) is working when you see (more pedantically, don't see) blind strips.
- Never give up.

Continue viewing or photographing for several seconds after the last vestiges of the solar image vanish over the horizon. The Novaya Zemlya mirage might be working in your favor, particularly in colder climes.

- The unique viewing of a green flash must be the sunset immediately after the total solar eclipse in Australia in December 2002. Many observers saw many of them, the weirdest being isolated patches where the Sun's crescent met the ground, whilst the larger portion was still well above.

The conditions met all the criteria I've set out here: a wide flat desert plain exposed to strong sunlight during the day and exceptionally clear skies as the Sun went down.

- Never forget to look out for green flashes with a setting Venus or Jupiter in your exceptionally clean skies (Figures 9.6, 9.7, 9.8, and 9.9).

Now that you've seen the green flash, always turn your view through 180° and try to spot the earth's shadow rising in the East.

Continue observing the sunset well into twilight for late crepuscular rays.

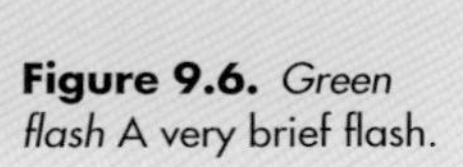

Figure 9.6. *Green flash* A very brief flash.

Figure 9.7. *Mirage* Mirage and unusual sky color did not lead to an anticipated green flash.

Figure 9.8. *Green flash, 2006* Mirage spread out the sun's upper edge into a line with green ends. A normal sighting in most places.

Figure 9.9. *Green flash, 2006* Some green coloration appears well out from the solar upper edge.

9.6 Red Sky at Night

The atmospheric conditions for "Shepherd's Delight" can be used as a weather guide so much more dependably than the morning's "Warning." However, never rely on it, either.

9.6.1 Mechanism

The proverb works, when it does, when a wet weather front is just moving away into the Eastern horizon. These conditions often lead to a trailing edge of cirrus and similar cloud formations that hang in the sky as a screen for many of the other glows.

- The conditions are not right when the front has completely gone and the much drier and colder air from the following cold front is upon us. That either leads to "Clearing up showers" or a totally clean and dry atmosphere, often devoid of excess water vapor that causes a red sunset.

A good red sky is often, not inevitably, seen after a long period of good weather. In that sense, at least it can be a predictor of continuing good skies.

9.6.2 What to Observe

Enjoy.

Keep that weather eye open for all the other glows as we prepare for nightfall. As I say, enjoy this classic red dusk in Figure 9.10 as you relax and wait.

Figure 9.10. *Red dusk toward America* Sea mist mixed with cloud generate a classic red sunset.

9.7 Later Glows at Dusk

All the dusk glows have been seen in their dawn or daytime counterparts and only need a brief description here.

Here are some tips to help you get ready.

9.7.1 Comets

If one of these comes along and is bright enough to be seen with the naked eye, then we're in for treat.

Always be prepared to see one of these and photograph it. The TV news flashes often mention these rare events long before the papers.

- All you need is an old-fashioned camera using a standard lens and reasonably fast film. That's never a bad tip to pass on. Don't throw your old equipment away, keep it for an emergency such as this.
- A good comet, that's one easily seen without optical aid, will cover quite a lot of sky and that will need a wide-angle lens if you intend making the picture into an art form with foreground interest.

All modern cameras and digital videos should be able to cope with this glow without any special preparation. Video lenses tend to be very fast and usable in extremely dark twilights. However, the cheaper end of the range will have a grim resolution and a modest zoom range.

- As in all things photographic seen only in the sky, a manual focus lens is almost essential, unless lucky, and it locks on quickly.
- A tripod is another essential prop if you want to get decent pictures. Some of the fainter comets test the timing function on longer exposures.

9.7.2 Earth Shadow

Always expect to see this in the Eastern sky immediately after staying out to see the slightly earlier sunset events. This sighting takes place in the very early stages of Civil twilight.

- The shadow should be visible any day the Eastern sky is free from low cloud. Whenever the cloud cover is down to wispy stuff, you might be lucky, usually not. These borderline conditions apply whenever the sky is set up for solar pillars and crepuscular rays.
- By far the most reliable skies are those immediately after a green flash has been seen. The air is generally extremely clean and that allows the shadow to be projected right across the sky to what is still a quite bright blue/green backdrop.

Under these conditions, particularly when the day has been warm and sunny, there is sufficient water vapor in the atmosphere to act as our screen and show the shadow to best advantage. Conditions become perfect when the air cools rapidly in the East and some incipient mist forms, not enough to hinder us seeing the shadow, merely enhancing the contrast effect.

- Good displays of earth shadow should be seen for at least 15 min in most countries before the inevitable happens and puffy clouds form and break up the complete arc across the Eastern horizon.
- The purple arc is always visible, of course, and in really clear skies the reddening to orange upper border never ceases to amaze the uninitiated. Pictures a minute apart show the distinct coloration (Figures 9.11 and 9.12).

Time lapse shots or videos remind me of the scene in Fantasia as the shadow envelopes us as we gambol below.

Figure 9.11. *Earth shadow over France, 2006* Immediately opposite to sunset.

Figure 9.12. *Earth shadow over United Kingdom, 2006* At 90°, to the North.

9.7.3 Dusty Skies

Because the advent of dust(s) into the atmosphere is so unpredictable, and rarely announced in telecasts, try the Bishop's Ring test during the day, heeding the warnings in Chapter 8.

Really bad dust pollution can generate all sorts of funny sky colors and effects. That's the only good thing to be said for them. However, long-lived volcanic dust has produced at least one good picture for me, the cover for a book on eclipses (Figures 8.3, 9.13, 9.14, 9.15, and 9.16).

Figure 9.13. *Bishop's Ring* The characteristic arc and colors around the setting sun.

Figure 9.14. *Bishop's Ring later* Changes in coloration and intensity happen very rapidly as light levels fall.

Figure 9.15. *Bishop's Ring, stars, and rhinos* Nearly dark and only the red remains as the animals emerge.

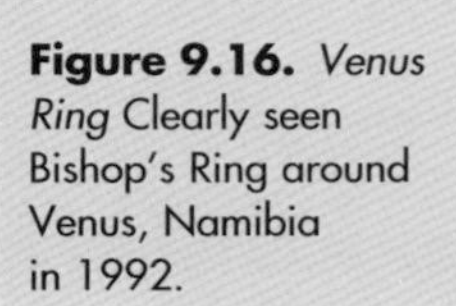

Figure 9.16. *Venus Ring* Clearly seen Bishop's Ring around Venus, Namibia in 1992.

9.7.4 Twilights

Never ignore these natural glows, no two are ever quite the same.

As astronomical twilight ends, the fainter stars become visible at the zenith, finally all over. That signals the night watch.

9.7.5 Zodiacal Light

We know night is fully upon us as the Zodiacal light makes its appearance.

We have come full circle in our day with this glow last seen in our dawn (Chapter 7). The Night Shift begins.

CHAPTER TEN

Night Lights

10.1 Introduction

All the bright lights seen in the sky at night deserve to be given much more attention than this briefest of introductions. Please consult a book devoted solely to pure astronomy for a more detailed review.

In this chapter, I've selected the more obvious glows that are clearly seen (or should be) without optical aid. This criterion is the primary difference from what's normally considered "real" astronomy.

You can use a telescope or binoculars by all means, but many of these glows cover such a vast area of sky and are so fast moving that a "proper" telescope becomes next to useless. Only a "Mark 1 eyeball" is needed, two if you still have them and haven't burnt them out by not heeding my advice on the Sun's dangers by day.

Although the Zodiacal light can only be seen under truly dark night sky conditions, it is really only a dawn or dusk phenomenon for most of us and is better thought of as such. Because most readers live in a northern continent, and the really difficult time to view is the early hours of the morning, I've chosen to go into the details in Chapter 7.

The Zodiacal light in reality does span the whole sky but that's extremely hard to spot because light reflection and perspective effects stop this from becoming very apparent again until the sky immediately opposite the Sun. This is briefly discussed in Chapter 7.

Gegenschein is discussed in more detail in the following section and is a true glow of the night—but that's not "nightglow," which is a separate topic.

In all the previous chapters, the effects of air pollution can be dire enough. At night the effects are devastating when combined with light pollution (Chapters 2 and 3).

We now pass onto other familiar and not so familiar sightings.

10.2 Gegenschein

Truly this is a glow only of the darkest and clearest skies. Something never seen and unlikely to be anywhere near to a town.

10.2.1 Mechanism

The name is derived from the German word for "counter glow," which is a very good description of the place in the sky, which is always on the opposite side of the sky from the Sun, the antisolar point.

Like the Zodiacal light, it has an origin in sunlight directly reflected back, backscattering, from the myriad of dust particles orbiting the Sun in the ecliptic plane. This time, however, it's specifically named from light on the opposite side of the sky or, more pedantically, from the dust cloud that is far away in the solar system.

- The difficulty in seeing it arises from the basic geometry involved. Sunlight has to traverse the solar system twice, there and back from wherever the dust is at that time, and only from particles of the right size, shape, and orientation to act as reflectors.
- Then, because the average reflectivity of solar system debris is around 5%, the Gegenschein has around the same reflectivity as the Zodiacal light's average. All in all it is a very difficult thing to see unless conditions are close to perfect. Otherwise it merges into the background or permanent aurora.
- This phenomenon has an analog in the daytime sky in glories (Chapter 8).

The mechanism is precisely the same in that sunlight is reflected back from particles (in this terrestrial case from water droplets or ice crystal) in clouds. The main difference from a glory is that a variable proportion of the light is polarized due to ions and molecules in the general backscattering medium in space. In either case both glows gain some polarization from reflections on ice crystals, to make this topic an extremely complex one to study. The mechanism is easiest seen in dew at dawn as the Heiligenschein (Chapter 7).

Quite unlike these daylight glows, the Gegenschein never shows up the dark central portion caused by the Earth's shadow. As with the glow seen from a very high aircraft above the cloud base, the shadow merges into a minute dot, then is swamped by the overall lighting and diffraction effects. The Earth is simply much too far away from the dust clouds in space to show any appreciable central shadow or spot.

10.2.2 What to Observe

Because of this premium on clear skies, the best place to see it is from a high mountaintop in or close to the equator. The modern observatories in the Andes fit this criterion extremely well, where the amount of water vapor in the atmosphere is also lower than most other places. Elsewhere in the Northern Hemisphere, the best period to see it is in the fall when cold, dry, and clean air starts to sweep in from Polar Regions. The glow then appears in convenient parts of the sky in late evening.

- From mid-September the glow is seen immediately below, making an approximate equilateral triangle with the "square of Pegasus" 's Markab and Algenib, centered in or close to the indistinct Pisces circlet of stars. Moving in a counter direction to the Sun at the same rate of about a degree a day, the glow moves to the Taurus region's characteristic asterism by the end of November.

Sometime in that period you should get a moonless night and clear skies. Leave the viewing any later and the Milky Way starts to intrude and overwhelm this faint glow, and you must wait until February and March for another window of opportunity.

During the next months of the year the thing appears much too low in the sky to guarantee good conditions, and completely hopeless the further you go north into mid-summer.

- All through this slow passage, across the sky you'll also be rewarded by the best seeing conditions for M31 and M33 seen strung out in a line at equal spacing in the early days of November. M31, Figure 10.1, will be of comparable size to the Gegenschein but considerably brighter, whereas M33 will be of similar brightness and much smaller in size. If you cannot see M33 with your unaided eyesight (with glasses if *normally* worn, not for the occasion) then the chances of seeing the Gegenschein that night are very slim indeed.
- *That's the other criterion for seeing it, perfect night vision.*
- For anyone in the Southern Hemisphere the windows are much shorter due to the more extensive Milky Way. Work on the basis of April and August either side of the winter solstice as your best chance.

At other times of the year and at other sites, finding the Gegenschein becomes much more of a gamble without these convenient markers. Guess where the Sun is behind you and hope for the best is the only advice I can give.

10.2.3 Some History

Because the glow has the appearance of a nebula and similar size of a comet, it was often mistaken for one. Barnard is usually given the credit for the first of these mistakes, saying at this best time of year, on 4 October 1883, that he'd found "...an immense comet." On that day it was at its closest to Algenib. He checked with the best comet hunter available in USA and got a reply from Lewis

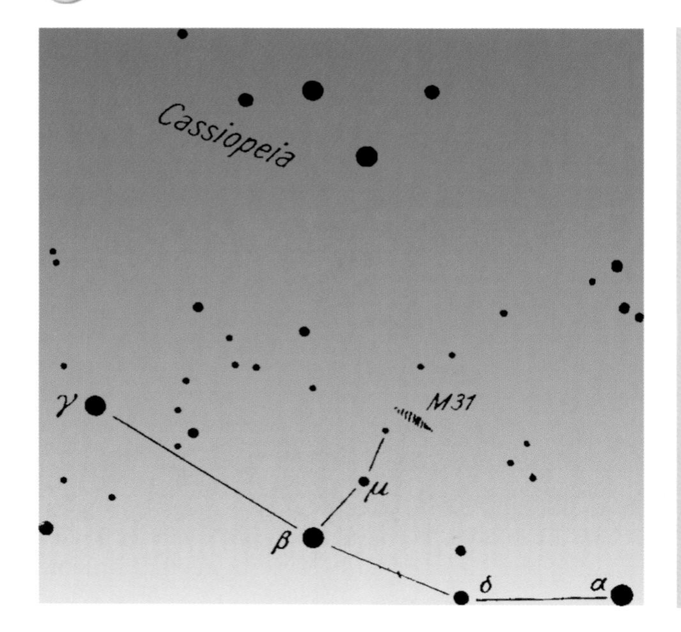

Figure 10.1. *M31 in Andromeda* Sitting below the "W" of Cassiopeia a good eyesight test in modest skies.

Swift (of Swift Tuttle fame) and the correct identification. Even with his keen eyesight and interest in fuzzy patches, Lewis had to admit he'd never seen the Gegenschein.

We'll never know who first saw or identified this strange glow in the night sky as there must have been many in Andean cultures whose constellations were based on the Milky Way and similar glows. Under those clear skies it couldn't have gone unnoticed, particularly due to its movement. Pliny in Classical Roman times, two millennia ago, commented on comets moving and could well have seen and not differentiated it. In modern times the first written record seems to be down to Alexander von Humbolt two centuries ago in 1803, during one of his voyages in that part of the globe, traveling in and around the Pacific off Peru. The Humbolt current gets its name from the same person.

In a sense Pliny might well, unwittingly, have hit on the glow's correct explanation. It does derive from the same cause as a comet's tail-sunlight reflected from the dust cloud emitted from a cometary nucleus plus ionization from the gas ejected. Much of the solar system dust cloud in the ecliptic plane will come from debris left behind comets, joining material from primordial times. He clearly distinguishes the things we now call nebulae as *fixed* and the moving ones as *comets*. The Gegenschein moves slower and always in a dark sky. The proximity to M31/33 has to be fortuitous in this description.

10.2.4 What to Check

Because of its rarity due to this dependence on a time of year and sky clarity plus the premium on superb night vision, this glow is really one to savor.

Much of the science associated with it has come on in leaps and bounds due to space probes and then the advent of CCD technology but to see it with your own eyes still leaves some things to check. Once seen or recorded in your camera, these other mysteries of size, shape, and position remain.

10.2.4.1 Size Perhaps the first surprise is the same Barnard experienced-the sheer size of it. Not a small point, as many expect, nor as wide as M31 which will be more than a full Moon in these clear skies, but much larger. Under the best conditions the camera records it out to about 20° across and the eye is quite capable of following it out to a similar extent. How far can you trace it? Those with perfect eyes and skies might well be able to trace the extent right round the horizon(s), and that will confirm the glow as being the same thing as Zodiacal light. Particularly this is true as dawn approaches and that sister glow intensifies.

10.2.4.2 Shape

- Because this is merely another part of the Zodiacal light, do not expect it to be perfectly circular, at least all the time.
- Oval sightings are very normal and the brightest patch clearly isn't always central.

10.2.4.3 Position

- The biggest surprise of all is that although a genuine phenomenon due to the material in the ecliptic orbit, the Gegenschein won't always appear where it should be. Quite why is still a matter of debate, and almost certainly due to the irregular or patchy distribution of the stuff in orbit, plus our own position within that.
- The central or brighter patch may be some degrees above or below the true ecliptic plane and that also changes from night to night. This variability does seem to have an origin in the same reason why the many meteor streams exhibit dramatic peaks of intensity. In this case the sunlight is reflected back from a particularly dense band of matter above or below the main rubble streams in space.

You will always know you're dealing with the Gegenschein if the object reaches the meridian and its highest elevation at local (true) midnight. Because we only see it in good conditions, checking its movement throughout the whole night for this obvious feature is a relatively easy thing to do.

10.2.4.4 Solar Activity

- The solar wind has a major part to play in exciting the gaseous or nonparticulate matter in the depths of space, and more and more observers are recording variations in intensity at times of heightened solar activity.
- Keep a lookout whenever a solar storm upsets power lines or communications networks, or an aurora is seen in unusually southern latitudes. Just hope that the aurorae don't interfere as they're being seen much further south of late.
- The next sunspot cycle maximum is predicted some time after 2010.

10.3 Nightglow

There's always some confusion on this topic that really ought to be described as aurora. The confusion is hardly surprising, as we're not thinking of the very familiar "bright stuff" seen near the Earth's poles but still arising from the same or similar causes. This time the glow is just on the limits of visibility.

This distinction with that name isn't at all obvious and the better names for this glow are "permanent general aurora" or more prosaically as "sky glow." However, that name has stuck, so we're also stuck with it.

10.3.1 Mechanism

There isn't a single mechanism for this glow as it's a combination of many quite different factors. These factors change all the time, making a proper study exceedingly complex.

What is quite clear, from our interest, is the glow is generated wholly within the atmosphere itself.

The major contributors are direct light of the visible and invisible stars, Milky Way, galaxies (and the brighter planets when on show) together with the streamers of Zodiacal light.

This isn't the light itself but the effect it has in general, light scattered from particles and other rubbish up there. Think of this component as the same mechanism as daylight on cloudy days. Much fainter, minute but definitely there.

Regard this as the same source as direct sunlight, only this time it's direct starlight and reflected sunlight from outer space.

Both these light sources on a moonless and planet-free night are for all intents and purposes a constant. That generates 20–40% of the overall background. The origin is almost entirely within the troposphere, or weather layers. Depending on what particulates are about, that will either enhance or screen the nightglow, the much more important and variable component from the emissions above.

There are real emissions of light on the other hand, which are definitely from the luminescence of the Earth's atmosphere itself.

Strange as it may seem, the light comes from sunlight, but not now. During the day ultraviolet light acting on the gases in the atmosphere ionizes and dissociates them. Some of these gases are the normal air components, which form ozone and other unstable molecules, whilst others are derived entirely from artificial "additives" we put there.

This interaction is a separate debate on the ozone hole, not important here. During the night a recombination interaction of this cocktail occurs and the energy of ionization and dissociation absorbed during the day is reradiated.

That process is phosphorescence, best described as energy in in one form and out in another. Remove the energy source of illumination and the glow stops. Luminescence is the same sort of thing but the process goes on after the initial injection of energy, just as an alarm clock gradually fades when the room light goes off. Both these processes and simple chemical reactions all add to the background nightglow.

Note that this energy is a totally different mechanism from a true aurora, which is due to external particles. However, some of the more energetic radiation hitting the upper atmosphere becomes a part of this nightglow when not obviously a full-blown aurora. This is also direct solar radiation, but sent into space hours, days or even weeks earlier. This is composed mainly of protons or hydrogen atoms. This radiation bombardment reaches a crescendo at sunspot maxima along with our familiar bright aurora.

The emissions take place in all the upper atmospheric layers. The normally accepted bands are between 40 and 600 miles (70–1000 km). The contribution is changing all the time throughout the night, season and sunspot cycle. That will on a normal night vary from 60 to 80% of the total. The major factor in all this is your geographic position.

Some of the highest energy cosmic ray particles from deepest space add very special radiation to the nightglow, known as Czerenkov radiation. I mentioned this in Chapter 5 as something Astronauts found very worrying. Figure 10.2 is a terrestrial example of that glow you can see at any nuclear power holding pond where the radiation levels emanating from the fuel rods are high enough to see. It is this blue glow, but extremely faint, that is our final contribution to a true nightglow.

Czerenkov's contribution to nightglow is actually not a simple process as Section 5.4.2 might suggest. Any particle traveling at near light speed (sometimes referred to as "relativistic" speed) which is faster than light speed in the atmosphere will do. At lower reaches, the air is denser and the relativistic light speed lower. That means solar radiation can contribute much of that background glow close to ground level, particularly at times of enhanced activity, flares for instance. But that's highly unlikely at night. The "flight time" for particles is a matter of minutes; we only see the slower particles with lesser energy caught up in the magnetic field. Relativistic particles in space stop for nothing.

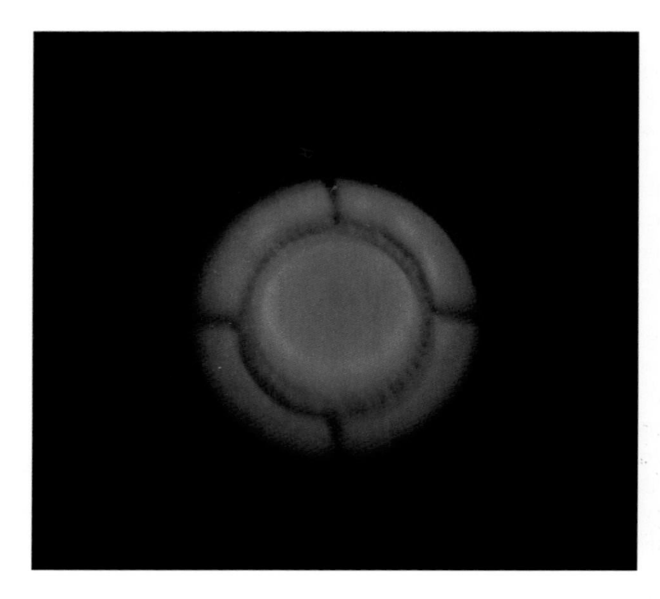

Figure 10.2. *Czerenkov radiation* Particles traveling faster than light (in water) generate this characteristic blue light, seen here in a nuclear reactor. A handheld photograph! Cosmic rays create this radiation in the upper atmosphere and that becomes part of the nightglow.

The astronaut's problem is the more fascinating and worrying aspect of this radiation. Sometimes a cosmic ray has super energy, running into unbelievable ranges, and that generates a pulse of light as soon as it reaches the upper atmosphere. However, because of the super energy levels, the particles can disintegrate air atoms on their way down and you get secondary pulses of light as these secondary disintegration fragments also move at relativistic speeds. By studying these Czerenkov light pulses as the different showers materialize on the way down, a whole new area of fundamental physics and astronomy is possible.

Figure 10.3 is just one of many major research projects now under way, the study of cosmic ray showers. Some amazing telescopes have been used in the past to capture this intriguing pulsing sky glow. Hopefully we won't ever see bright Czerenkov radiation in either guise. If we do, that could well be the last thing we ever see. It means that a supernova or similar catastrophic event nearby has generated enough relativistic cosmic rays to bathe the Earth in a deadly glow of the most extreme energies possible. There is a lot of evidence that's a definite cause of major life extinction and/or genetic diversity in the Earth's past history. Definitely not on our list of or even:

10.3.2 What to Observe

The direct light from stars and so on remains extremely constant unless Venus is around. Except in the tropics, this Venus effect is never measurable against the twilight. Jupiter can scatter a lot of light as Mars does every 2 years.

The gaseous emissions create a weak uniform and constant illumination of the night sky on any moonless night. This is never seen anywhere near to scattered light from cities. The generally accepted brightness from this source works out

Figure 10.3. *Czerenkov scope La Palma* A telescope to monitor the elusive flashes of light in the sky, now finished and operational.

to be about two to four (about) fifth magnitude stars per square degree of the sky. That's minute, but obviously there. This emission glow predominates over the natural-light, one from stars and so on, and will be 60–80% of what you see.

The spectrum is as we expect it. In addition to a continuous spectrum, there are emission lines and bands from molecular oxygen (e.g. exactly the same as the green and red lines of the Northern Lights), nitrogen, most of which are also the same as the Northern Lights, and sodium (the famous D-lines). The closer you are to the sea, the greater this contribution from sodium, which means you can't get away from that stuff.

The spectrum and the total intensity of the night sky glow aren't constant everywhere or all the time at a fixed point. The variations seem to depend somewhat on the activity of the Sun. That introduces two more variables strongly affected by the Sun's activity.

1. The ultraviolet radiation responsible for the daytime ionization and chemical changes of the atmospheric gases is never constant and the spectrum (that's the energy content) doesn't always match the Sun's output in other ways. Visible light doesn't change much, whereas the ultraviolet can change 1,000-fold or more in minutes.
2. The solar wind creating direct ionization at night.

All of these ionization effects are also highly dependent on the Earth's magnetic field in the atmosphere and that in its turn is influenced by solar activity.

Hardly surprising it's difficult to sort out this lot. To monitor the nightglow until recently wasn't feasible except in major research laboratories. Today you can follow the progress with ease with any of the Nightscopes on general sale.

Our next nightglow is even more intriguing.

10.4 Luminous Bands

Whether or not this should be included here is a moot point. I've chosen to because it doesn't fit into any of the regular aurora or nightglow categories. It seems to be a special case of both, very rarely seen.

10.4.1 Mechanism

These bands are reported in the past to *only* occur immediately after a major meteor shower or storm. This could well be true if the storm occurs in daylight, so nobody ever sees it, or the shower hasn't been picked up by radar because the particles are too small to register.

I find that theory increasingly untenable because modern radar is more than capable of picking up swarms of insects. Meteors whether of the stony or nonmetallic variety should be an easy target for radar specifically designed to pick up "stealth" aircraft.

The theory, nevertheless, runs like this:

- The meteorite particles are much too small to create visible meteors and are either simply burned up in the upper atmosphere or stopped extremely quickly. That dust settles extremely slowly through the atmosphere and creates a haze. The effect is not to block starlight, more to enhance it by diffraction effects.

A considerable amount of debris from outer space does reach us in this way, and there can be no doubt that much of this comes in discrete orbits, associated or not with known comets and asteroids. Many theories in the past associated this mechanism with unusual rain patterns, for instance.

This mechanism must remain on the table until radar or something new in technology settles the argument.

- The upper reaches of the atmosphere where this is reputed to take place lie in the same region as many of the other transient glows. This is up to about 75miles/120 km.

This higher reach makes it somewhat different from terrestrial dust coming up from below. Terrestrial dust generates the normal sky glow.

10.4.2 What to Observe

There's little to add except to be on the look out for major meteor showers and plot the skyglow with your Nightscope over the next few days, that is to say clear nights.

A considerable increase in the skyglow illumination in the atmosphere follows this phenomenon. That alone seems to confirm something particulate on the increase up there when the solar effects are nonoperative.

As hinted at, it's also been linked to increased rain and a whole host of other fanciful disasters. The rain theory is quite widely held.

That it's a real effect and not a mistaken isolated faint aurora, Noctilucent cloud or similar is absolutely definite from the huge number of observations in recent decades.

Do not band this together with any other nightglow, as the appearance is distinctively that of a barrow strip or band.

10.5 Noctilucent Clouds

Noctilucent clouds are one of a group cataloged under "Mesopheric Clouds". Although as you see them for the first time they have many similarities in appearance, these are quite definitely a totally different mechanism from Nacreous (also called Mother of Pearl) clouds (Chapter 7).

The name comes from the Latin "Night Shining," a perfect description.

They are such a fascinating topic that I'm now beginning to be convinced they play a hitherto unsuspected role or even a vital ingredient in our understanding of the changes going on in the climate above us and the weather we experience down below. To do proper justice to them, the whole book merits devotion to the clouds alone, if only to get to grips with their descriptive catalog types.

An earlier book has already been devoted in this series, and should be consulted for the essentials, now unfortunately out of print (*Noctilucent Clouds* by M. Gadsden and W. Schroder). I can only hope to whet your appetite and entice you to delve further, go out, and observe, and then draw your own conclusions from some of the suggested observations.

At one time these were thought to be the same as luminous bands and have the same origin in their formation from meteorite debris in the upper atmosphere. The jury remains out on their precise origin, but here are a few ideas.

10.5.1 Mechanism

Whatever these clouds are, they're definitely on the increase.

- As normally seen they only occur in latitudes close to the poles in the few weeks or months around the summer solstice. The skies are never truly dark at this time of year and it begs the question whether or not they should be considered a true nightglow.
- The clouds are never seen except in the nautical twilight when the Sun is between 6 and 12° below the horizon (Chapter 6). Because they're too faint to be seen in the later/brighter stages of twilight, all authorities catalog them as a night event. This is becoming truer as the Sun can be as much as 16° down and still create a display.
- At this time of the year the Mesosphere, hence the name of Mesospheric clouds, drops to its lowest annual temperature. Normal chemical reactions are at a minimum at this temperature–125 to 130 °C.
- The years around sunspot minimum appear to favor a good display. That is probably due to the low incidence of aurorae at a time when the Sun's ultraviolet and X-ray radiation are at a minimum. On really good nights both glows can be seen together.
- To be seen at all the sunlight must illuminate a critical layer that is normally confined to a narrow band bordering the upper mesosphere, although mainly in with the lower Thermosphere. That places them above 50miles/80 km when normally seen.
- The layer's height is markedly different at the poles. In the North it's some 6% lower than in the South in direct relationship with the amount of sunlight energy reaching them due to the different distance from the Sun in an elliptical orbit. As a result the Southern summer is a week shorter.
- At this height the amount of water vapor is minimal which seems to rule out water condensing into ice on meteorite debris.
- Clouds when seen overhead and to the South are relatively faint, indicating that light is scattered forward from the Sun in the North. As the displays seem to be showing further South each year, they risk "vanishing" (i.e. to say diminishing in apparent brightness) from their conventional abode in the far North.
- The geometry from all this suggests that an appearance is only possible in a symmetry around local true midnight and the summer solstice. Until 2006, this

was normally the case but the recent sightings have become more excitingly erratic.
- If we discount meteorite debris and water vapor (ice) from down below that doesn't leave many options to explain what the clouds are. At that height the air is too cold and rarefied to support most chemical reactions but with strong sunlight we return to a normal day at sunset/sunrise when the radiation is almost as high as at noon.

The culprit seems to be methane. Well not exactly methane itself. It seems that methane is broken down in the strong sunlight to give water and carbon dioxide as end products. Burning, if you will. This is the normal mechanism removing methane from the atmosphere as it filters upward from natural gas, coal mining and other industrial processes, and paddy fields and cattle.

The evidence is now much stronger that the water vapor formed crystallizes immediately into minute ice crystals and it's these that we see glowing in the dark.

Whether or not this is a major contributor to the "greenhouse effect" remains to be verified. The clouds are some 15% brighter than from as little as 20 years ago, which indicates that this boundary layer is actually cooling. This paradox confirms the extra cooling at these altitudes during high summer.

Methane is a much more powerful agent in this process and that gas is increasing much more significantly than carbon dioxide.

The Aeronomy of Ice satellite was launched in September of 2006 with the hope to resolve as many of these conundrums as possible. ICE was fitted with instrumentation to:

- Measure meteor particles down to 50 nm. This is in the same particle range as smoke from forest fires down below and should resolve that issue;
- Image the clouds from above, to correlate with images from ground level;
- Measure the temperature. In the course of time that should resolve whether or not the mesosphere does or has cooled significantly in recent years.

Maybe these noctilucent clouds are the harbinger of our downfall?

10.5.2 What to Observe

Very much like a camel, extremely difficult to describe on paper, but once seen instantly recognized and recognizable. There really isn't anything else about to get confused (Figure 10.4).

- Invariably seen against a golden to orange twilight glow to the pole's horizon, observers these days see them down to 50° latitudes, or even below. They should be expected in any part of the sky.
- The clouds have every appearance of some variety of cirrus, but of a ghostly iridescent blue to greenish hue.
- Their shape and size varies over such a wide range that the catalogs of the sub-classifications are quite bewildering. Best to accept that any extensive shape as being one if it fits the iridescent nature and looks wispy. A crude method but effective.

Figure 10.4.
Noctilucent clouds
Credit—John Murrell.

- To say that noctilucent clouds are on the move is a bit of an understatement, as these really have the skids under them. Speeds in excess of 90 to more than 500 mph (150–800 km/h) are their normal range. Because of their high altitude averaging 50 miles (80 km), the apparent motion you see belies the hurricane speed. With these two pieces of information, try plotting the motion and calculate their speed at night.
- Photography with a conventional camera is extremely simple with a tripod and a timed exposure.

400 ISO film is readily available and with a standard 50 mm fl lens of f1.8 a second's exposure should capture enough for a pretty picture. Stopping down to f2.8 is usually recommended to enhance foreground interest, in which case exposures up to 4 s are needed.

Always use slide film if possible, as commercial processing with prints is quite unpredictable.

Digital and video are a lot easier to handle and give an adequate resolution, with better sensitivity to subtle coloration(s).

- Whenever the opportunity occurs, always use plenty of film or memory and try to photograph at regular intervals. That timing data can then be used to make sensible calculations of the motion and speeds involved.
- Make best estimates of the angular coverage. A finger outstretched approximates to a degree. Relate your estimates to the pictures.
- Always record days when nothing is seen. That negative information is as vital to know as all positive sightings.
- Finally, never forget to record the direction first seen and subsequent motion, then note the time at each stage. In most countries, some form of Daylight Saving applies and that must be noted or corrected to UT accordingly.

Do not underestimate the importance of these clouds and do enjoy them during their brief life in summer. With AIM now in orbit, this is one area where amateur observation can be critical in our understanding of the future environment. You might be the only person to see something in a brief (i.e. to say troposphere) cloud break. Summer weather can be notoriously fickle and not all will be as clear as in 2006.

Our next group of glows is more of a treat in dark winter nights, the aurora.

10.6 Aurora

A true aurora is a term usually reserved for a full-blown colored display close to the Earth's magnetic poles.

In the Northern region, the nightglow is commonly called the Northern Lights, or technically the Aurora Borealis and has a counterpart in the Southern Lights or Aurora Australis. In either case, the glows do normally occur only in latitudes close to the poles, more specifically to the magnetic poles.

The basic mechanism and understanding of these familiar nightglows is extremely well understood and few readers will be unaware of what they look like. This brief summary is intended to encourage trips to see one as the solar sunspot cycle rises to a maximum in the next few years.

10.6.1 Mechanism

The mechanism is extremely similar to nightglow in two of the previous sections.

- Particles from outer space impinge on the gases in the upper atmosphere and cause a whole cascade of electrical and chemical changes. Most, but not all of these changes are ionization effects and the atmosphere is stimulated into a glow that's extremely familiar as a red light in a neon light or bulb. Neon is much too small a component in the atmosphere to react like this and can be discounted.

The red light is seen when the much more common oxygen is activated in the highly rarefied upper reaches around 250 miles (400 km) high. The red light seen at that higher altitude is normally well into the red part of the spectrum (630 or 636 nm). It is extremely intense and by far the most obvious in a good display.

Under these conditions the impinging radiation is so energetic that the oxygen is split into two and the radiation comes from oxygen atoms.

At extremely low bombardment rates we can regard this as one of the (effectively) invisible components in the nightglow background.

- Under fiercer bombardment the glowing region moves rapidly down to the thermosphere boundary around 55–65 miles (90–100 km), sometimes even lower to 50 miles (80 km), which is the region of the noctilucent clouds.

On rare occasions both can be seen together as happened in 2006.

This rapid rate of progress downward is one of the explanations for the rapid movements and "curtaining" seen in a good display.

As this glowing region moves downward the same oxygen atoms in the relatively denser atmosphere start to radiate in green light. This green glow is also very intense and is seen at 578 nm.

In a normal display the incoming particles may be highly energetic but in small numbers, so that all we see is an indistinct greenish glowing patch somewhere in this, the lower part of the sky. That's most people's experience.

- Another and much more fascinating auroral glow is often seen as a purple (some see this as violet) band or patch. That normally occurs at the top of the green and red lot and can easily be missed.

When that happens the bombardment from space is usually extremely intense as the radiation comes from a much greater height, up to 650 miles (1000 km), and is from excited nitrogen molecules, some broken in half. The color is 391 and 428 nm, on the borderline for color detection by many.

To add to the general confusion, the much fainter purple regions, due to their much greater height, are often just within a sunlit band, and that can turn the whole display a lot bluer.

How the aurora is formed by these incoming radiations is a major research topic going on all the time in a huge number of countries. Knowing when an auroral storm is imminent can mean the difference between life and death when power lines go down.

The basic mechanism arises from particles emitted from the Sun (the solar wind) reaching the magnetic field of the Earth.

In normal circumstances, this magnetic field can and does act as a shield and the particles are either diverted around it or reflected completely (Figure 10.5).

All hell breaks loose when one or more factors occur.

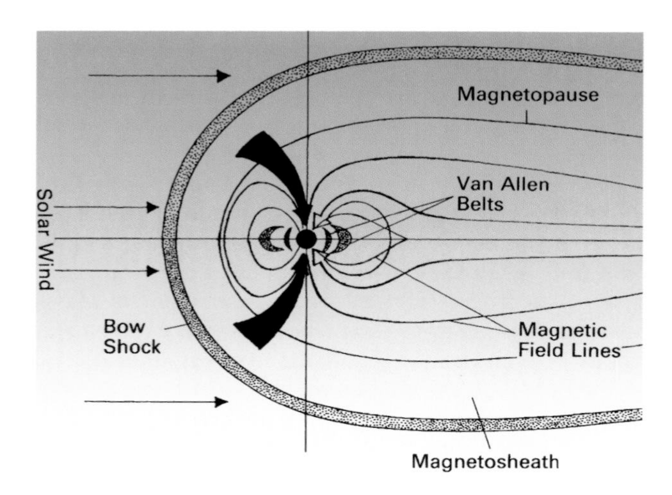

Figure 10.5. *Earth's magnetic field* Acting very much like a bar magnet in the space around us, the magnetic field is subjected to huge fluctuations when the sun ejects charged particles during a big storm. The Earth's field can be destroyed completely on occasions, leading to massive power and electronics failure. See NASA web pictures for the most up-to-date information, in colour. Credit—Neil Bone.

- When a solar flare on the Sun's surface ejects more than the usual amount of its atmosphere, the corona, it then becomes part of the normal solar wind, then arrives with much more energy than usual, and is able to punch through the protective magnetosheath. With enough energy some will reach the magnetopause, where the Sun and Earth's magnetic fields cancel out. Whilst not an entirely true analogy, the solar wind particles behave like an electric current and go to Earth. Just as in a dynamo, they are caught in the much stronger magnetic field at the magnetic poles and try to reach Earth as they spiral down. They never reach the ground (except in very unusual cases) and are captured by the atmosphere to create wonderful colors.

Enough enters the Earth's atmosphere to create the aurora on most nights at all times of the solar cycle, but rarely seen far from the magnetic poles.

The solar wind is variable, just as with any other wind, and that shows up as the pretty color and movement changes.

- Occasionally, fortunately very occasionally, there is a major eruption or flare on the Sun. In this case the particles are so energetic and fast moving that they reach the Earth in a matter of hours rather than days and with so much energy they bash straight through the protective magnetic field as though it is not there.

The nett effect is to totally wipe out the Earth's magnetic field, reverse it completely in huge tracts, or even create huge currents within the Earth's surface. All and any of these features will completely disrupt communications networks, power lines, and many forms of delicate electronic gear.

These are the famous geomagnetic storms. When they occur, extremely dramatic aurorae are seen and these can venture way out of their normal polar territory to almost anywhere on Earth.

Occasionally, the aurora shows evidence of hydrogen alpha, beta, and other radiation. That implies that some extremely energetic radiation is getting through to disrupt atmospheric water and other molecules.

The really worrying factor, not taken into account in most climate models is the way the Earth's magnetic field is rapidly declining to a total field reversal sometime in the next millennium.

That process started its rapid decline at a time coincident with the start of the Industrial Revolution in mid-nineteenth century, coincident with the putative start of global warming.

That rapidly declining magnetic field will lead to ever-more common aurora displays and more damage to ground-based structures as the radiation from space is able to creep ever lower. Already there have been satellite failures in the South Atlantic, close to the magnetic South Pole.

10.6.2 What to Observe

The sight is so familiar that it hardly merits description. An excellent simulation is provided by this PC's Media Player I'm listening to as I type this section. To see what I mean, go to Windows Media Player (OK, I know you're using iPod!),

Figure 10.6. *Aurora* A sequence taken at short intervals. The display was bright enough to read a newspaper.

tab "View", then "visualisations", then "Musical Colours " and choose "Aurora". Very good, just a trifle too red as normally seen, but it does capture the whiteness often seen close to the top. That's a genuine color most of the time. Or much more likely simply your color deficiency in the blue.

Once past the wonderment stage of seeing an aurora for the first time, there are several things to observe.

- The first is to get ready. That's particularly relevant the further south you live as guaranteeing a view isn't as certain as in Alaska. One good tool to start with is any of the Aurora watch websites. The other is to make a magnetometer or even simpler to closely monitor a compass. If it veers off to an odd direction for no apparent reason and stays there for longer than a few seconds, check the sky outside at night for an aurora.
- Second is to make scientific measurements of what you see. Consult your local Astronomical Society for advice and to whom to send results.

Be aware that the classifications are extremely complex and detailed, but that's part of the fun. There are at least eight descriptions for the overall form, similarly for the symbols and structure, before you get onto conditions, brightness, and color, some dozens to work to.

Let's pass quickly to photography.

10.6.3 Photography

The brighter the display, the easier this is.

The best advice, as film is cheap and the experience isn't, is to take as many pictures as you can at any one sighting. It might be the only one like it for years.

With a 35 mm camera the faster films record quicker but suffer from loss of color saturation. As with most other nightglows, use slide film rather than print if processing in the high street, the reverse if doing your own.

The best shots normally need a wide-angle lens, but the faster the better.

As a rule of thumb for the brighter displays try 3–5 s at f2.8 with 400 ISO film.

With digital and video, consult the instruction manual. They should all cope well.

For the 1992 display in England, I was able to read a paper by the glow and check my camera settings without a torch! A T-70 camera set on auto-exposure coped superbly as you can see (Figure 10.6).

Aurorae are the true miracle glows of the night as seen by their magical displays of movement and color.

Son et lumiere au naturel. Some people have even recorded sounds and music, and that's almost certainly true if the rapid magnetic changes affect us in a way some have suggested.

10.7 The Milky Way

For many of us the Milky Way is just one of those things we take for granted as a sight to be seen on any clear night. Few will not know of its existence, although how many do actually go out and look for it? Precious few I guess in these hectic days. Less as time goes on with light and atmospheric pollution becoming more prevalent.

This is one subject that puzzled the ancients as to its true nature and place in the scheme of things. For the civilizations in the Andes, this was less of a puzzle as the glowing band of light across the sky was so much more obvious in their rarefied air at high altitude. Their constellations included the Milky Way as the main theme and once you've seen it under their conditions, it's easy to see why. The dark patches matter, and the other stars fill in the gaps. In the Old World, the stars took precedence and the glow was a bit of a nuisance to a tidy mind.

10.7.1 Mechanism

This subtitle is a bit out of place here as we now know there are uncountable stars in our own galaxy. Best estimates say around the hundred thousand million marks, usually shortened to 100 billion. This is too big a number to comprehend until fairly recently but the advent of computers and "Banana Republic" economies have made billion a common figure to bandy about.

The best way of thinking of it is 1 s in 30 years.

The stars aren't evenly distributed and until the end of the nineteenth century nobody had the faintest idea how they fitted together, nor how they related to the other galaxies. We now know that the Milky Way is a mere dot in the universe and even with the galaxies in our immediate neighborhood lumped together and called the Local Group, we still remain a dot in the Universe.

Our nearest neighbor of comparable size is the Andromeda nebula, known as M31 that I've referred to it as a good guide for the Gegenschein (Chapter 10).

It's not until one goes to the Southern Hemisphere that the Milky Way can be appreciated in a full glow. The central regions in Sagittarius are spectacular. All the more so when our nearest galaxies, the Magellanic Clouds, join them.

The reader is referred to an astronomical encyclopedia for further vital statistics.

10.7.2 What to Observe

For most of us in towns, there isn't any option-you can't. When I grew up the Milky Way was visible from almost anywhere that smog didn't affect. As late as the 1970s it was still possible to see it clearly in suburbia. By the 1990s street and security lighting put paid to it on all but the clearest of nights.

So, what do you look for at a good site?

- First and foremost, just enjoy the sight whilst you can.
- The next step is to use a modest pair of binoculars and take a good time scanning and see how many stars you can actually see.
- If you have a telescope, start with a low power (or wide-angle) eyepiece and repeat that exercise.
- Finally, use the highest power you can and do the job again. Guaranteed to put spots before the eyes.

This is just fun, but without it nobody can gain any conception of the sheer magic in appreciating the environment and placing our worries in context.

Assuming it's not possible to travel to a good site in summer to see the central portion in Sagittarius, here are some other ideas.

In late summer to early fall, the constellation Cygnus (The Swan) Figure 10.7 is nearly overhead at a convenient hour, and a nice, rich part of the Milky Way goes right through that. At that elevation it should be visible in modestly light-polluted places.

- The test of your seeing conditions is easy to try. The brightest star is Deneb (which means "tail" in Arabic) with the three "wing" stars in a line immediately in front and some distance ahead is Albireo at the "head." If your sight is any good, you should be able to see a distinct dark band cutting into the Milky Way, right behind the "wings" in front of the "tail" and an even more distinct band appearing to cut the constellation off completely right behind the tail.

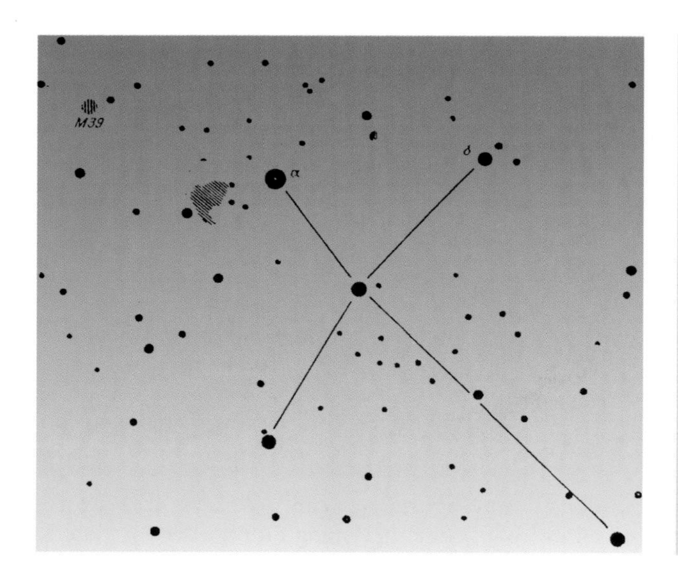

Figure 10.7. *North America nebula* An iconic light in the sky that's easily picked out with good red-sensitive eyes and crystal clear conditions. A "must see" light.

This distinctly dark patch is often referred to as the "Coal Sack" of the North. If you live in the South, the genuine "Coal Sack" will be all too familiar. That part of the sky is by far my favorite, Figure 10.8.

This is an optical effect as the dark band in Cygnus often doesn't quite make it but at those low light levels, close enough. Poor seeing conditions and all this description will mean nothing since the Milky Way will be suffused with its own glow from lighting below.

- If your site passes that test, then you're lucky and can now try for a gorgeous glow, the North America nebula. In a perfect site this is clearly seen as a "smudge" almost touching Deneb. As it's reddish, that does place a premium on good color vision, of course.

With a modest pair of binoculars, the general shape of this glow should be immediately apparent and why it has that name.

- Obtaining a photograph is a much more difficult job without a proper tripod and motor drive. If you can do that, then the Sagittarius part of the Milky Way will repay your time and effort with some stunning detail and ever-more fainter glows. The constellation is absolutely stuffed with interesting glows and small groups of stars that look like a glow. These are known as clusters (Figure 10.9).

In the Southern Hemisphere, you're spoilt for choice in glows to seek and enjoy. My second favorite is another cluster with a myriad of brilliant spots making up a distinct fuzzy patch. That's Omega Centauri. Any star atlas will cite all the statistics on the number of stars present—mind boggling stuff.

The Andromeda nebula has already been mentioned several times. If anyone asks how far you can see, that glow is the object to quote. Light takes 2.4 million years to reach us, and we all know how fast light travels.

The Andromeda nebula's distance makes it the furthest glow we can see with the unaided eye—a true sister to our own Milky Way.

Figure 10.8. *Coal sack* Arguably the nicest group of lights in the sky, the southern cross region. The coal sack is that dark patch at the left-hand edge, best seen by averted vision as an "absence" of light below the cross, *top left corner*.

Figure 10.9. *Saggitarius clouds and 'travelers'* The brightest part of the Milky Way is increasingly difficult to photograph without some form of unwanted light from aircraft or satellites. The overexposed spot is a natural "traveler," Jupiter.

10.8 Other "Fuzzies"

Comet hunting has always been a mainstream activity for astronomers and has led to many new scientific discoveries. However, because the objects are invariably faint when first spotted, a huge number of false alarms were made in the early days.

10.8.1 Mechanism

Charles Messier, born in France in 1730 was also concerned by the problem and set out to catalog "fuzzies" that could be mistaken for a faint comet. That led to the famous Messier Catalogue of 1774 and two shorter supplements in 1780 and 1781. It's not generally realized that he wasn't the only contributor or that later additions were made.

Today the list runs to 109 items out of the final 110, as 102's identification is still doubtful. That list is extremely helpful in the Northern Hemisphere and a total loss for the South since it made no attempt to cover that rich part of the sky.

The list includes many genuine fuzzies that are readily mistaken for a comet, such as the Andromeda nebula, No. 31, but the list also includes some surprises such as open clusters of stars. These loose collections are not seen as a fuzzy patch to anyone with normal eyesight. The Pleiades (M45) were known as distinct stars from very early in our human history. Other objects and galaxies (not known as such at the time) are so faint as to make the whole exercise more than worthwhile.

Many other catalogs were drawn up over the years and the Webb Society is just one of many.

The latest addition that's caught on like wildfire in America is the 1993 Caldwell Catalogue, initiated by Patrick Caldwell Moore, now Sir Patrick. This lists another matching number of 109 objects missed from Messier's Catalogue, including the important Double Cluster in Perseus and Southern objects. That list is very practical in content.

10.8.2 What to Observe

No list of objects to observe can ever be complete without the inclusion of the Messier Marathon. This is a firm favorite in America and readers are hereby encouraged to check local web sites and Astronomical Societies for the next one near you. Join in and have fun. Being in a group like that pays off when trying to identify the really faint ones that can only be found with telescopes at a fixed observatory.

Most take place around the equinoxes in March and September.

If you can, add the Caldwell list to your observing run or make that a separate Marathon.

I don't need to list the whole gamut here, as that's so well catered for by web sites and textbooks. My suggestions for a starter have already been mentioned in M31, M33, and Omega Centauri.

Don't take my word for it, just get out there and see for yourself how many "fuzzies" you can spot before turning your binoculars on the next lot, the ones that benefit from a little optical aid.

Start with these:

- NGC 7000 the North American nebula in Cygnus
- 47 Tucanae, a nice globular cluster
- NGC 2237-9 the Rosette nebula

Then go back to the Scutum Cloud as part of the Milky Way and see if you can find M11, the Wild Duck cluster, my favorite in the Northern skies. Check to see if you agree with me that it does look like a flock of ducks in flight.

When hunting for comets and meteors, our next quest, never miss checking the Messiers, Webbs, and Caldwells.

10.9 Comets

Surely the gem of all nightglows, comets have evoked awe and fear in equal amounts throughout the ages.

10.9.1 Mechanism

Their main value in our study of lights in the sky is so obvious as to need little further comment. Sir Edmund Halley is credited with the first scientific study of them, hugely aided by the work of Newton on gravity and laws of motion. In 1682 he finally identified the orbit of one bright recurring one which bears his name, the famous Halley's comet.

Halley is justly famous for that groundbreaking research but not fully appreciated for the calculations of no less than 24 cometary orbits. That study is the foundation of all modern work.

Halley's comet was a great disappointment in its last return in 1986 due to the way the orbit's appearance from Earth stopped the tail from making its usual spectacular display. The tail pointed at or away from us instead of spreading out in a normal fan fashion. The science gained from the close encounter of the Giotto probe is still being studied, and probes to other comets on their way or planned.

- Some comets make extremely regular returns like Enke's which is seen every 3 years and 115 days. Since its discovery in 1818, it's been round more than 50 times by now. You need proper equipment to see this regular glow in the sky.

There are a huge number of these regular (short period) comets and much modern research is now devoted to confirming which are genuine comets, as distinct from asteroids. Both can coexist it seems, the main distinguishing feature being whether or not a tail forms on close approach to the Sun.

- Other comets come from way outside the solar system, starting in the Oort cloud. Named after the Dutch astronomer, this is a region of space out to around a third of the distance to the nearest star where that gravitational field starts to take over.

These come in very fast and pass extremely close to the Sun and are often the brighter. A good one in recent years was Hyakutake (C/1996 B2), which moved so fast across the sky that you could almost see it go. The tail spread out almost across the whole sky (see Figure 2.3).

10.9.2 What to Observe

Be prepared at any time to go out and see one. Hyakutake was a classic. I was on an ordinary visit to La Palma's observatories when it appeared from nowhere and was seen right overhead within days. The picture in Figure 2.3 was taken on the hotel roof by the simple expedient of laying the camera on its back. See notes below on keeping a spare camera ready for just such an event.

The picture also illustrates how, even in such a tightly controlled environment, light pollution (Chapter 2) is still a problem. The clouds are brightly lit.

Comets are always so newsworthy that the TV and papers will be a start. Some comets like Hale Bopp were visible all night for a time that extended for months (Figures 10.10 and 10.11)

Figure 10.10. *Hale Bopp and Mercury* Mercury joins the comet to make a nice picture with twilight glow.

- Whenever a good one, that is to say one easily visible to an unaided eye, makes an appearance, do just that and go out and see it.
- Whenever a fainter one is announced in the scientific press, do the same and try to spot it with binoculars or a telescope. Your local Astronomical Society will normally be only too pleased to assist.

An ordinary domestic camera can capture all the brighter ones.

- For conventional photography you need the fastest wide-angle lens you can get (the standard lens as fitted is never a disappointment), a fast film, and an exposure as long as possible up to 30 s or more for the fainter ones. A tripod is an essential stability aid together with cable release.
- For digital cameras the same rules apply except that the CCDs record to their limit anyway. For video cameras, these can also be used in "night-mode" in most cases and many allow frame capture via a computer. Consult the instruction manuals for their advice.

Whatever you do the next time one appears, you must be prepared and ready. Always keep an old pair of binoculars handy and above all never throw away

Figure 10.11. *Hale Bopp* Not all comets show a distinct gas tail (blue) with the familiar dust (white to yellow). Star colors and star clusters are seen in this picture.

an old camera or video. All too often your main one is tied up for some other project and you'll kick yourself for not getting a unique picture.

Note: I'd written this homily and was about to put the text to bed for the publisher when an email landed. It announced that a new comet had been discovered and should be visible in 2 days' time...

This is Comet 2006 M4 Swan. That makes my point that some of these only appear at dawn or dusk, this one at dawn, and are definitely not nightglows or objects. It's scheduled to make unaided eye visibility in the evening sky a few days later.

It will be past history by the time you read this piece of Stop-Press. The excitement of comets is in tracking them and seeing whether or not they ever come up to brightness expectations or become just another disappointment.

Good comet glow hunting. Now to something a bit more downwardly mobile, meteors.

10.10 Meteors

Shooting stars as they're normally called can genuinely be called lights in the sky. Some of them are so bright that they really grab your attention. All the ancients record spectacular sightings and Biblical accounts perhaps the best known, with stars falling like rain.

10.10.1 Mechanism

Some definitions first.

A *meteor* is any bright light seen in the night sky that's caused by something from outer space burning up in the atmosphere.

- The conventional brightness range is anything from the detection limit in a big telescope, or way below anything you can see, to that of Venus. That's taken as -4 magnitude by convention.
- A *fireball* (also called a *bolide*) is anything brighter. Something that you can't fail to notice, as some are more than bright enough to be seen in full sunlight.

The characteristic of these is that most leave a glowing *train*, burst into *flares*, sparks, and fragments, or all of these. Some leave a *smoke trail* that can persist as long as an hour.

The train or trail may also glow and become another type of nightglow in its own right.

This residual glowing train will also swirl and change shape very rapidly as the strong winds at altitude take hold, to eventually disperse the dust.

- A *meteorite* is whatever's left from either definition when it hits the ground. Although some may be white hot, they've ceased to be a light in the sky and don't come within our interest here.

- A huge number of meteors arrive all the time but most of them are so small that they pass completely unnoticed. On any clear night you should be able to spot one or two of the brighter ones every hour. These are called *sporadics*.
- A *meteor shower* occurs when the number of sporadics increases dramatically and virtually all of those appear to come from a small point or area of the sky called the *radiant*. A *storm* occurs when the numbers seen reach "astronomical" proportions, normally in excess of several per minute.

The majority of meteors are incredibly minute and one bright enough to be seen is no more than a millimetre across. One heavy enough to reach the ground isn't much larger if it's traveling a lot slower.

The brightest meteors are traveling at a speed of tens of miles (km) a second, which is much increased when the Earth is (a) orbiting round the Sun into stream coming the other way and (b) rotating into a stream coming straight at us, something that normally happens near dawn. Conversely, the approaching Earth and the relative speed drop to a very low point where ordinary gravity takes over as we catch up a slow stream.

This is the basis of the theory behind luminous bands (Section 10.4). The fastest are traveling at up to 45 miles (72 km) per second whereas the putative luminous bands come from free fall stuff that's not consumed when traveling below 7 miles (11 km) per second.

The amount of material used up in producing the light is very low, and only about 1% of the heat appears that way, together with some ionization of the atmosphere. That ionization trail allows the meteors to be picked up by radio and radar and be detected in daylight. Some of those showers only show up in daytime and are lost to us.

The faster meteors burn up at a great altitude, well above 70 miles (110 km). The great majority burns up in the 70–56 miles (110–90 km) region; the fireballs can and do go all the way down to ground level.

At the higher altitudes the meteors burn up in highly rarefied air and are hardly slowed in the process. The best analogy is one of the air molecules hitting the meteor like machine gun bullets, which heat to white heat in the collision energy and burn off the surface. Some gas is ionized in that energetic state.

Larger objects start to meet ordinary air resistance below about 30 miles (50 km) and by 6 miles (10 km) most will have slowed to normal gravitational free-fall speeds, just like a parachutist. That's a partial explanation why some of the larger objects seem to stop burning before hitting the ground.

However, the majorities reaching these lower altitudes are still traveling fast enough to experience the full force of air resistance and temperatures in excess of 3000 °C are reached as the air isn't able to move out of the way fast enough. You get a ramjet effect. Coupled with the huge incandescent air resistance on fragile meteors, the whole lot can disintegrate with a spectacular display of celestial fireworks. An accompanying thunderous noise is also quite common to make this a noisy glow.

The link with comets in the previous section is extremely strong and many of the brighter streams and showers have definite parentage in known comets.

Comet Halley is the best known, as it generates two meteor streams each year as the Earth's orbit intersects the comet's.

The Eta Aquarids radiant is less well known because it's so far South. At best it produces up to a meteor a minute sometime during 5–6 May. There isn't a definite peaking of activity in that time and several subpeaks are normal due to the age of the comet spreading out material each time it orbits the Sun.

The Orionids are much better known although only half as many will be seen around 20–22 October. What they lack in numbers is more than made up for in spectacular fireballs, many leaving quite persistent trains. Like the May apparition, several subpeaks are possible.

Of the two, the Orionids always impresses. The familiar Perseids in August come from Comet 109P/Swift-Tuttle and the best of the lot these days is the Geminid shower, peaking around 13–14 December with up to two a minute. These are slow moving and often leave a train. They come from an asteroid, 3200/Phaethon.

10.10.2 What to Observe

The joy of meteor watching is just that, the wonder of it all whenever a bright one catches your attention. Some of the Leonids in the early 2000s were bright enough to light up a room indoors. Many of these left luminous trains and smoke trails that persisted for ages (Figure 10.12).

Meteor watch parties are very popular, and readers are always welcome to your local one. Why? Well, this branch of astronomy is one of very few scientific

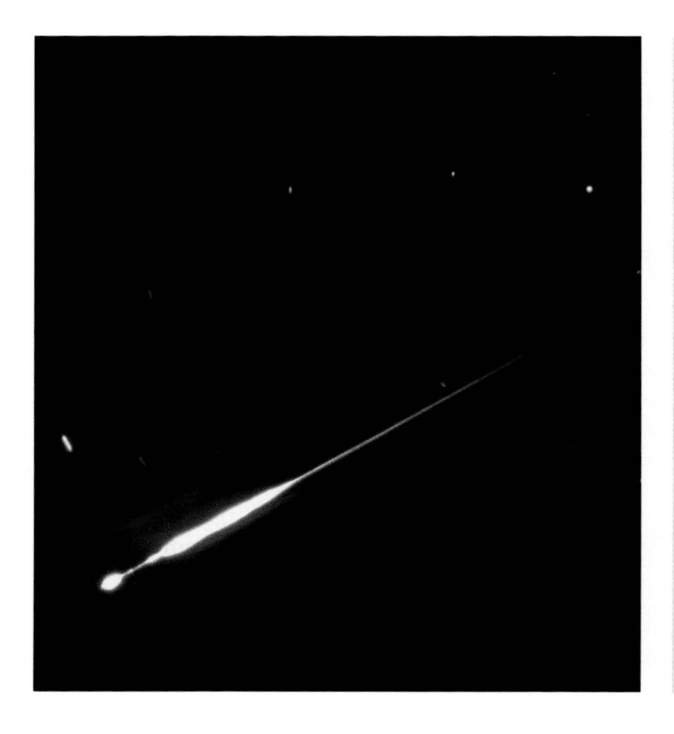

Figure 10.12. *Perseid* Not all meteors are as bright as this one, nor show flashes and a distinct train that persists.

pursuits where the true amateur can make a positive scientific contribution, and make a genuine discovery.

- The name of the game is to be keen enough to go out on any clear night and simply keep watch on any set region of the sky. (Do not be tempted to keep glancing around, the science is to monitor a set region.) Set aside periods in discrete hourly periods and just note how many meteors you see.

That simple. Send your results to a local Society, or better still get into groups of friends and ask one to be a recorder. How to go about this is readily available from Webs and Societies.

- The second useful bit of science is to be a bit more scientific and make notes of the meteor's characteristics—speed, brightness, color, and so on, and where you think it came from. Identifying new radiants is always a good way to get your name in the scientific journals.
- The best science comes from a truly scientific setup. You don't need any special equipment if you can't afford it but a simple camera, just fitted with a standard lens and any old film (you're not normally going to print from it, just to measure). That's another occasion when obsolete gear pays dividends in a new lease of life.

The most valuable task to undertake is to organize into small groups and see if you can triangulate and plot back to where the meteor's path goes. Even that simple equipment can identify a meteor stream and new ones.

A simple addition of a rotating shutter and it's then possible to plot the speed and really tie down the orbits.

Give these ideas a whirl and join the band of amateur scientists adding to original knowledge from the simplest of gear and the simplest of all the nightglows.

Become a bright spark yourself.

10.11 Stars

The most obvious glows in the night sky haven't been considered so far, the stars. There's a very good reason for this, apart from the obvious one that a full study of the many mechanisms working is pure astronomy from any normal textbook.

Stars do have a few surprises, best kept till last as here in this chapter as that might be the last thing we see!

10.11.1 What to Observe

10.11.1.1 Colors For anyone who's never taken the trouble to look through any optical aid, perhaps the biggest surprise is to find that most stars aren't white at all but are of every conceivable color and brightness (Figures 10.1, 10.7, 10.8, 10.13, 10.14 and 10.15).

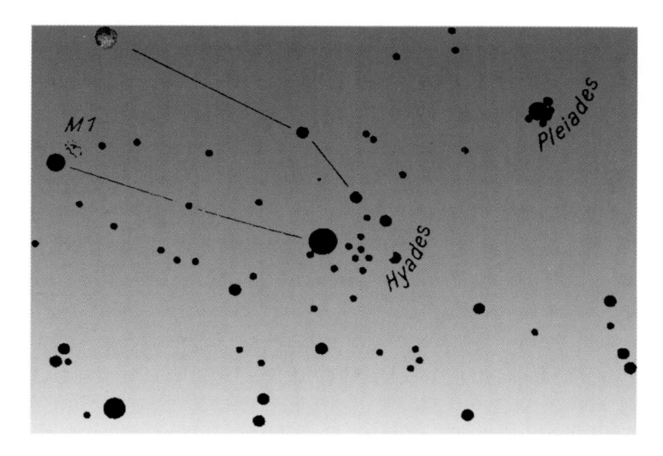

Figure 10.13. *M1 in Taurus* A real test of sky clarity and eyesight. Normally a light seen only with optical aid.

Take the time to check this out, and see how many colors you can identify. Yellows are extremely common and reds too, but the big surprise is to find some genuine white ones that can merge into distinct blues/azure. The color is a direct measure of the star's temperature, and we gain most of the information about them from a study of the color and spectrum that they emit.

I've already suggested looking at Albireo in the head of Cygnus. An absolute must is to turn some reasonably good binoculars on it and note that's not a single star, but a pair called *double stars*. These two are quite different in color, the brighter one is yellow while its fainter companion is distinct blue to green. The contrast is staggeringly beautiful, never to be forgotten.

Some apparent doubles aren't proper doubles, merely an effect of two in line-of-sight. That's not important for us as this is an enjoyment exercise.

Some genuine star systems contain more than two disparate star types and five is not rare. Once one multiple has been found, the hunt can be addictive.

10.11.1.2 Variables The next surprise is to find that many stars (including our Sun) aren't perfectly stable in their brightness—certainly not stable at all wavelengths or colors. Some of the variability has already been covered in a review of aurora and in other topics, where the solar output in the extreme ultraviolet is often from virtually nothing to major storms.

Our Sun is a genuine variable and has been steadily increasing its visible light output over recent decades, which is minute but quite measurable, but which takes quality instrumentation to detect. This factor is rarely taken into account in discussions on global warming (hence not considered at all by the more extreme environmental pressure groups). As far as we're concerned in this book, many of the marked increases in the rarer sky glows are entirely down to this variability.

It seems that all stars are variable to some extent. The difference from our Sun is that some other stars make it a (genuine) way of life and death.

Cepheids are a case in point. These stars have reached a point in their lives where the whole star swells and shrinks at regular intervals, and are known as *pulsating variables* as a result. This variability is so constant, and the light at

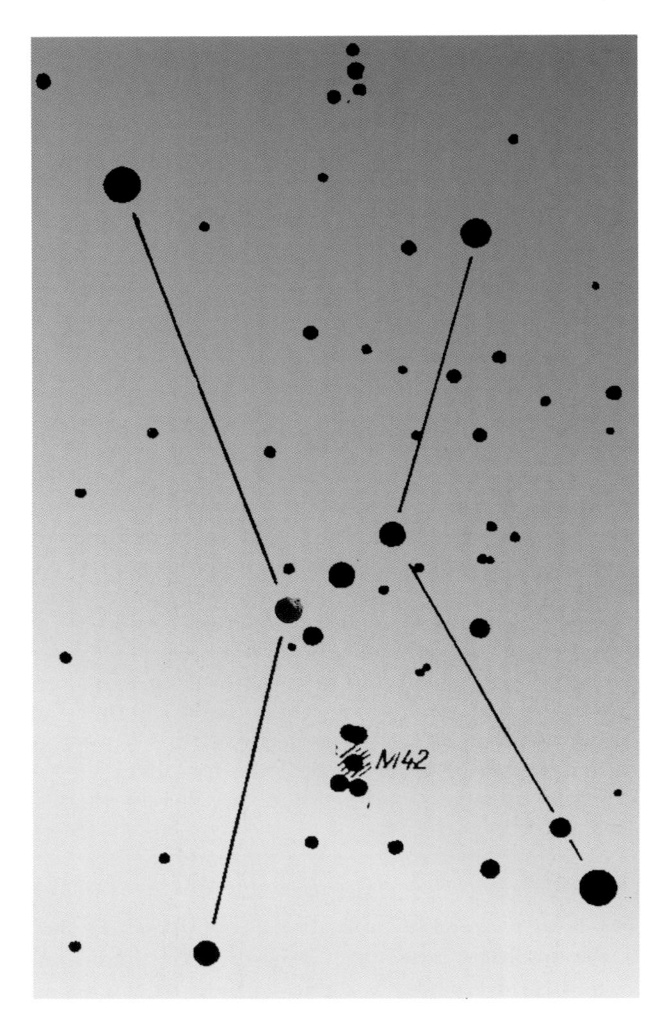

Figure 10.14. *M42 in Orion* It has to be a very poor sky, or eyesight, not to see this light in the sky.

each point in time so predictable, that these stars have become one of the main work tools of astronomy.

It was through a detailed study, about the distances to distant galaxies like the Andromeda nebula, that we now have a much better handle on the sizes and distances involved.

See if you can find in a star atlas the first of these "standard candles" used to measure our universe. It's Delta Cephei, not far from the better known Cygnus.

You'll be "doubly" rewarded when you find this sky glow because Delta Cephei is another double, easily split in modest binoculars. The brighter one is described

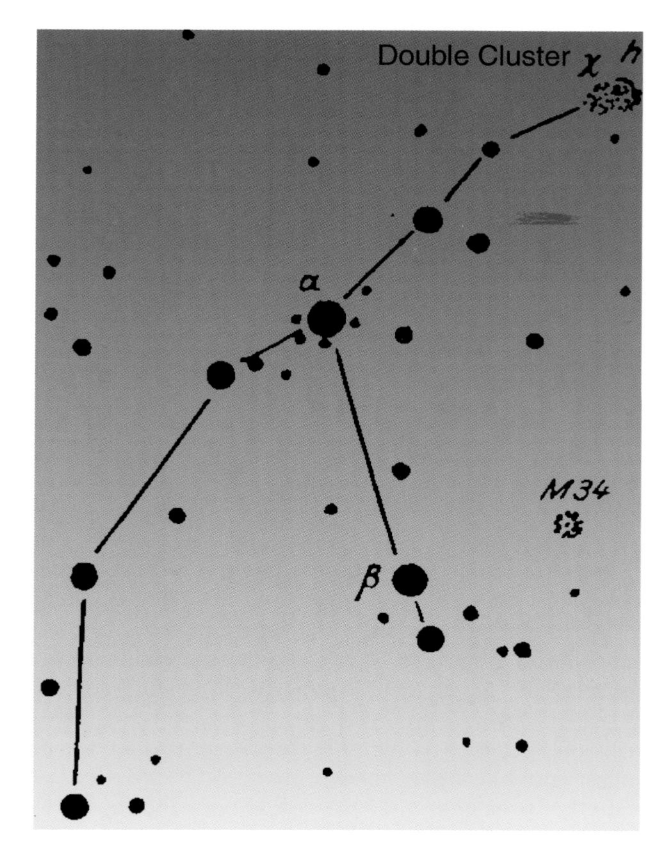

Figure 10.15. *Perseus double cluster* Most people see this as a single fuzzy light in the sky. Look again and see its true nature at the bottom of the inverted Y in –the sky. See Figure 10.11 for the colors appearing in photos (the patches at the end of the blue gas tail).

as a cream color, and the fainter sixth magnitude one, a little way off, is another blue to white specimen.

A few stars have been known from antiquity to change their brightness and one is infamous from its very name—Algol. That means "Devil's Head" in Arabic. First cataloged in recent times in 1669, Algol is normally the second brightest star in Perseus. See Figure 10.15 where Algol is β. It varies from a faint 3.5 magnitude to a very obvious 2.2.

In mythology that variation, every 2 days and 21 h, lent to the idea of a winking eye in the head of Medusa. Perseus is depicted as holding forth that severed head that still winks. We now know that the star is but one of a group whose variability is due to stars orbiting each other and causing partial eclipses. The star also lent its name to a whole class of *eclipsing variables*, but turned out not to be typical in being a triple. Even today, we still don't fully understand how that system operates.

There are many types of variable stars, but by a long straw the most important are glows of really apocalyptic proportions.

10.11.1.3 Novae and Supernovae

The precise mechanisms by which these stars explode have become one of the major driving forces in professional astronomy. They are so important from two viewpoints here.

1. We are only here because in the primeval past those exploding stars were the factories in which the elements forming our bodies were created.
2. If a really big one happens anywhere nearby, that'll be the last thing you see, as you become a glowing part of those stupendous sky glows. This is a proper nuclear explosion.

If the second comes to pass, even the whole Earth could be vaporized in an instant, as the energy released in a supernova can equal the whole light output of the host galaxy.

Many amateur astronomers spend their whole spare time looking for these stars, always hoping to catch one about to happen. Most of them are caught some time afterward.

Professional astronomers don't have the telescope time to do a survey and rely on amateurs to find them first. That collaboration is unique in the scientific world. Can you imagine an amateur brain surgeon being let loose to do the original work?

Then as each new specimen is identified, then confirmed, the full works are unleashed in observatories all over the world.

This is yet another area that you, the reader, can indulge in some genuine scientific research. Some in the UK with modest equipment have discovered well over 200 in the space of a few years. Mostly in galaxies that don't boast a name, just a catalogue number. However, these are still genuine glows in the sky that can be listed and studied as they evolve.

The research field is still open for you to have a go. Only by catching enough novae and supernovae at different stages can we hope to fully understand how they work, and from that get a handle on the universe as a whole and how we got here.

10.11.2 The Aftermath

A study of novae and supernovae isn't as daft as you might think. A very faint star is left behind, often shrouded in its own "ashes." Most of the good ones leave behind some fabulous nightglows in those ashes that are listed in the Messier, Webb, and Caldwell Catalogues. The classic one is the Crab nebula M1 (Figure 10.13).

M1 is the remnant(s) from a massive supernova seen in daylight in 1054. It's a huge gas cloud that's still expanding and that can be plotted quite easily from pictures taken at regular intervals. Tracing that movement back, we get the same 1054 start-date.

This remnant's glow is a real test of eyesight and quite impossible to spot without the essential conditions of a crystal clear sky. The essentially green color doesn't help identification for any one with color blindness. Make it a test for wherever you are. Photographs show many other colors.

Many of the more spectacular nightglows come from these nova and supernova remnants. My recommendations include ring nebulae. In approximate order of difficulty, try these two:

1. Messier 57 The original Ring nebula in Lyra.
2. Caldwell 39 The Eskimo nebula in Gemini.

Pictures certainly confirm their names from the appearance, and as for the colors in modern faked (sorry, enhanced) pictures, it is a good pair to stick on a wall as decorations to admire.

Some very recent research and studies has been devoted to identifying and spotting what are known as gamma-ray bursters.

Some regions of space contain the most energetic events known to date. These are so cataclysmic that it seems the whole energy contained in a galaxy suddenly appears as gamma rays. They impinge on matter in space (maybe in our atmosphere) and show as a transient glow. These are faint but detectable and definitely lay claim to be another right glow.

As with supernovae, let's hope one doesn't go off too near. The snag seems to be that their identification, once confirmed as a real event, is on the increase.

Some scientists have suggested, with some evidence to back it up, that a few of the major animal extinctions in the past might, just, be due to a supernova nearby. Now, we must add gamma-ray bursters—one group of glows I do hope not to see up-close and personal.

10.11.3 Star Birth

We come full circle here as the material spewed into space after a supernova explosion rapidly condenses into dust and gas clouds that condense further and eventually coalesce into dense clouds. It's from these condensations that new stars are born after more gravitational attraction.

This has become another pioneering research field. Those exciting color pictures we see in newspapers and magazines often depict these star-forming regions, our final glow in this section.

The best known of these glows is the same famous Orion nebula M42, Figure 10.14, I now have so much difficulty seeing (Chapter 5). The full extent of this glow is around a degree. It is normally seen as green, but appears in photographs in a wide variety of reds and violets. This is entirely due to the way the eye performs at such low-light levels.

The whole region around Orion is a star-forming nursery as are many other huge areas around the sky.

Enjoy a trip round these as they're announced, before we come back to Earth for our final set of nightglows.

10.12 Other Nightglows

Most of the other glows to be seen at night are mere night variants of glows seen at other times of the day, better thought of as "anytime" events.

In this category is the entire halo and similar effects seen with the Sun, but this time seen with the Moon. Moon rays and crepuscular rays are rarely seen, but I've been on the lookout since starting this research and find them surprisingly common. Perhaps not so surprising as the weather conditions at night is the same as during daytime.

Rainbows are also seen at night by moonlight. Lightning is also an another 24-h a-day sight.

10.12.1 Eclipses

For a long time, the ancients regarded these in awe, as the Moon seemed to vanish from its normal brilliant glow and turn a terribly threatening blood red. Sometimes the reddening was so deep that the Moon almost vanished completely, which means that this light in the sky is a true variable, even a negative one.

We now know that the reddening is entirely due to the weather conditions at the sunrise and sunset rims around the Earth. If the sky is almost devoid of cloud, the amount of light reaching the eclipsed Moon is as great as it can get, and the Moon remains surprisingly bright. However, if there is a lot of cloud cover, then little light filters through, and the Moon can vanish almost completely.

This dire condition, to the ancients, also happens whenever there is a lot of smoke or volcanic dust around and under those skies worldwide; the Moon can go a surprisingly wide range of colors, including dark grey. That's one explanation of "once in a blue moon" as the event is so rare.

Normally in the year or two after a bad volcano, as in 1991 with Pinatubo, the Moon went a lurid red and almost vanished. Figure 10.16 shows how stars became easily visible.

Figure 10.16. *Lunar eclipse with stars* When the eclipse is total, stars not normally seen close by appear with this red light in the sky.

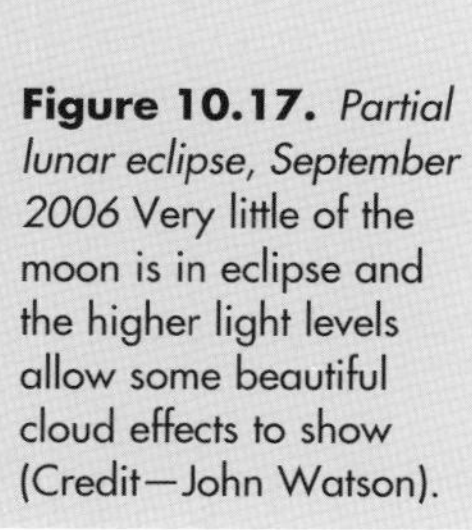

Figure 10.17. *Partial lunar eclipse, September 2006* Very little of the moon is in eclipse and the higher light levels allow some beautiful cloud effects to show (Credit—John Watson).

Much more common are partial lunar eclipses where only part of the disk is covered and a lot of moonlight remains to illuminate clouds to make a secondary light in the sky. John Watson's picture taken in September 2006 shows how a little imagination can generate a truly artistic picture (Figure 10.17).

10.12.2 Volcanoes

The volcanic effects on eclipses have already been mentioned above. Their weird secondary effect of generating electrical storms, lightning, is shown in Figure 13.2. In this section the light in the sky is entirely due to the eruption itself.

Little more need to be said, for one never wants to be too close whenever it goes off (Figure 3.11). We're long overdue for a big (Figure 10.18).

And now for something completely different, of all the many possibilities that haven't been considered already, perhaps the most dramatic are the unexpected sightings of satellites and other bits of debris up there. These do pop up out of nowhere, as if by magic.

10.12.3 Satellites

Nobody of my generation will ever forget the beginning of the space age when Sputnik went up. We all went out and tried to spot it but the most memorable sightings came later on when the space race got into top gear in the late 1960s. Today, we tend to take no notice at all of what's up there—familiarity shouldn't breed contempt.

10.12.4 What to Observe

On any clear night it should be possible to spot at least one satellite in motion. These are sometimes difficult to distinguish from highflying aircraft whose navigation lights are not obviously flashing. The best advice I can give is very straightforward.

Figure 10.18. *Stromboli* Once described as the "Lighthouse of the Ancient World," Stromboli's light in the night sky continues to operate every few minutes after thousands of years.

Aircraft should:

- Show navigation lights at sometime.
- Keep a more or less constant brightness when closest.
- Not deviate from a straight and narrow path.

Follow tracks that are familiar during the day (at least on the busier routes).

On the other hand, satellites:

- Are never colored as there is no navigation lighting
- Frequently demonstrate amazing changes in brightness, usually due to tumbling motion and sunlight striking new faces.
- To the unaided eye often appear to swerve from side to side rather than dead straight. An optical effect in our eyes/brains has nothing to focus on that far away.
- Follow set paths, more often than not due N/S.
- Vanish suddenly as they go into Earth's shadow.

Keeping an eye out for these satellites can be scientifically of value.

The much more dramatic glows that seem to appear as if by magic are created by the considerably larger structures already functioning up there or under construction.

Consult the media and Webs for the latest data on the Iridium, and larger communications satellites. With Space Shuttle now back in service as I write, the Space Station will be much more obvious in future and is always memorable each time I see it. That does materialize by magic and vanish just as mysteriously, and cannot be mistaken for anything else from its brightness and characteristics listed above.

At last, this is one glow that can be entirely predictable to a published schedule.

Watch this "space" as they say.

CHAPTER ELEVEN

Halos

The moon in haloes hides her head

This saying was already mentioned when we looked at the likelihood of the weather turning nasty after a red dawn (Chapter 7). And that's our first main problem as we almost arrive at the end of our tour of lights in the sky. What has been described in antiquity as a halo might not be one at all, but a corona or any one of the many other possibilities.

Halos and arcs, with corona, can occur at any time during the 24 hours, and appear quite unexpectedly. Contrary to common belief, they aren't rare. A major reason for not spotting them is the same for any unexpected sighting; nobody takes the trouble to look. The other factor is the sheer size of some of these phenomena. They can take over the whole sky, literally.

A complete review of all the possible permutations and combinations will take another considerable book devoted solely to halos and arcs. Here we'll only tackle the more obvious cases.

Some have already been separated out, as they're brighter and more common on average, and these normally arise from ice crystals in the plate form. The glows in this chapter are normally seen when the predominant ice crystal shape is the other variety.

True halos and arcs do not occur just in bright sunlight as many consider but as frequently in moonlight. It is another example of the simple question of relative brightness that stops the lunar halos and arcs from being more obvious. Sun dogs for instance in Chapter 8 have their night analogues in moon dogs.

The conditions for seeing moondogs, whose scientific name is *paraselenae*, are precisely the same as the daytime analogue. Whether or not you see at the same time a weak 22° halo is entirely dependent on the ice crystal structure in the cirrus cloud layers, not light intensity. Moon dogs, because the light source is so much weaker, are only regularly seen close to full moon, not because they don't happen.

A very minor problem in a review of this type is the spelling. The archaic version has an "e" as in the quote at the start of this chapter, while the modern spelling does not.

This summary outlines the more common features and some rare variants to seek out that we've not considered yet.

11.1 Mechanism

The number of possible halo permutations seems endless, and no single individual living at a fixed site can ever expect to see more than a sample. This is a taster to inspire further study of these finer points.

All the genuine halos and arcs we see are not a feature of liquid water droplets as in a rainbow or glory, but are due to the light transmitted through or reflected from ice crystals. Furthermore, unlike the more dramatically colored lights, they aren't seen away from the Sun on the other side of the sky, but rather as a feature around it. Somewhere we don't normally gaze at.

Or they can be seen around the Moon, of course, and all the mechanisms apply equally to that light source, just considerably rarer to spot.

All are formed and give all the brighter halos from the same ice crystals identified as hexagonal prism crystals (often referred to as "like a pencil"). For simplicity I'll call them needles (this is a chemist's description), as the ends' shapes don't enter into our study of the first and most obvious total halo (Figure 11.1).

These ice crystals are also responsible for the solar and lunar pillars (Chapter 7) and many of our other lights in the sky, and can also give us the more exotic halos seen from time to time.

The only criterion needed to see anything is a cirrus or similar thin cloud formation in front of the Sun that contains these ice crystals in abundance and

Figure 11.1. *Halo crystal shapes* Ice forming from atmospheric moisture creates these two main crystal shapes. How light interacts with them leads to a bewildering but a delightful range of lights in the sky.

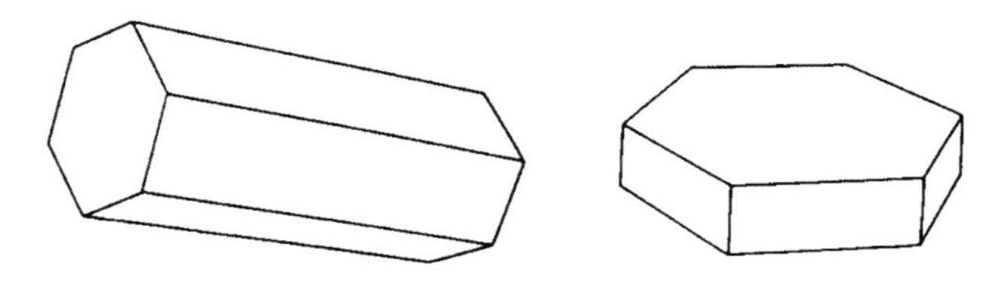

in similar populations. As a normal cloud of this type always occurs above 3 miles and up to 6 miles (5–10 km), even in the tropics, the chances are almost guaranteed to be cold enough to form ice. Three miles (5 km) is about the height of “the snows of Kilimanjaro” in such a tropical region.

The rapidity of formation appears to be one factor deciding which crystal form predominates. The rest is down to how the crystals are displayed in front of us.

The plates are easiest to understand as we can regard them as minute leaves falling from a tree. Most of the times they are flat and fall by sideslipping so that the overall effect is the same—they appear on average to remain horizontally flat. With the high winds often associated at those altitudes, the crystals don’t fall at all but are carried along flat as that’s with least air resistance. Whatever the cause, when the situation is thus, we can see the reflection mode of the solar pillars (Chapter 7).

Halos and arcs on the other hand are an effect of light refracted through, not reflected from the needles. And here’s where our explanations get complicated with these hexagonal crystals. Even now the total explanation is not fully understood, because to get our halo we need ice crystals to be randomly orientated, and not perfectly aligned as with the plates for pillars. And that’s the unknown piece of science why that should occur. Winds and other factors ought to sieve or sift into more regular patterns. Whenever that does happen, we can experience a much wider range of halos and arcs.

11.2 22° Halo

The first step is more obvious as light passes more or less straight through the body as though that were a simple ice prism (Figure 11.2).

It doesn’t matter how the crystals tumble or rotate as long as a majority is seen lengthways and horizontal at the same time. The sum or net effect is to create a complete circle of light from the curtain of crystals in a cloud in front of us. For a really good display, most needles do take up this state and are predominantly horizontal, the closer the better and rotating around that long axis only.

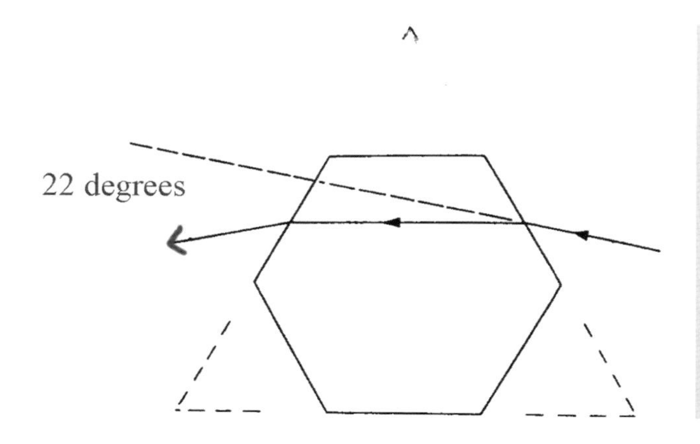

Figure 11.2. *Main halo’s light path* Think of the minute ice crystal needles (viewed here through the ends) as a prism. Then it’s easier to see how the light is refracted through it to our eyes to form a spectrum, hence the halo and the umpteen varieties of arcs have colors.

As with all the other refraction phenomena, light is concentrated into a tight beam or caustic as in Figure 11.2, and, in this case, the most obvious color is red as the least diverted or dispersed. We find a constant angle for that red light at about 21.7°, and for other colors not much different and just slightly wider.

This angular closeness lends its name to that bright annulus as the 22° halo. It is also called a *circumscribed halo*, but that is often confused with the merged arcs I describe later.

All colors overlap to some extent and the annulus or halo is never very brightly colored or intense until we reach the horizontal points where sun dogs occur.

Sun dogs always appear at the absolutely horizontal points to the Sun. Anything different is entirely due to a perspective effect, or an irregular horizon, or even poor balance, maybe too much alcohol consumption.

Whenever color is seen at all, the tight inner portion will be reddish, merging into oranges and yellows to blue or grey.

It is rare to see sun dogs with a complete halo as these are formed by light passing through the other common ice crystal shape, the plates we saw in Chapter 8, and both types don't always coexist.

The more normal situation is a long-lived display which at its most intense gives an appearance as though of a "hole in the sky," which is quite obviously so in these pictures in Figures 4.3 and 11.10.

11.3 The 46° Halo

If all the theoretical ray tracings are carried out, 22° isn't the end of the story as light can be dispersed all the way out to 50°. Pictures rarely show this; rather the effect is to enhance the visual impact as a dark hole in the sky. However, it is

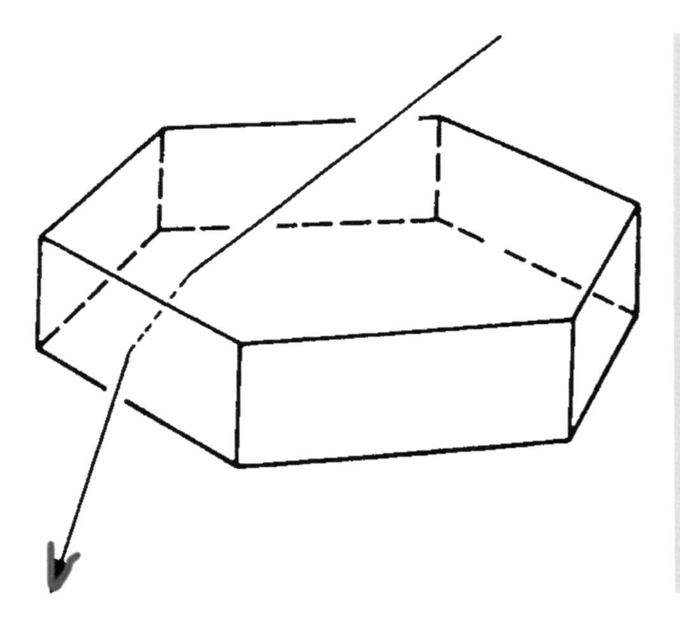

Figure 11.3. *46° halo* The secondary halo outside the more common one is formed from light passing (refracted) through the other ice crystal form, from above. The conditions are much more stringent, and the sun must be high in the sky.

sometimes possible to see an outer ring or annulus of light at 46°. That's not the end or maximum deviation, rather another caustic occurring at the analogous position to the 22° example when dealing with a 90° prism of ice instead of our regular 60° prism (Figure 11.3).

Our 22° annulus is our minimum deviation or caustic. In very rare cases the sun dogs' tails from other crystals go right out to an appreciable portion of the 46° halo.

The first complication is the height of the Sun in the sky. A full halo can, obviously, only occur whenever the Sun is at least at a 22° elevation. This is not rare at all, but because of our familiar factor of lack of observation most halos of this type are seen close to sunrise and sunset, when the eyes do not need to look up.

An explanation of all the other annuli and arcs depends on a much more complex set of refractions through our needles or plates, or one of the other rarer crystal situation(s) coming next.

11.4 Bright Arcs Type 1

Whenever the ice needles aren't random in our cloud but much more regularly aligned more or less horizontally, a brighter patch can occur above and below the main circle. That patch is often so intense that's all one sees.

This group of arcs is referred to as a *column arc* due to this crystal orientation, as though in piled up columns. The crystals do seem to be stabilized in this way due to a rapid rotation in two directions:

1. Along the needle (that is to say in an almost perfect horizontal plane) whilst also rotating and
2. In a plane at right angles (that is to say in an almost perfectly vertical direction) relative to our view.

When our crystals are like this, a column is in fact sometimes seen when the Sun is reflected (not refracted) from the same sides of these crystals. This is something that's best seen when the Sun is largely dimmed at the horizon, very much as with a regular solar pillar.

- The arc above is called an *upper tangent arc.*
- The one below is not surprisingly known as the *lower tangent arc.*
- As the Sun rises, these two arcs merge to form another brighter full halo just outside or merging with our 22° version.

In fact that elevation has to be 29° to gain proper clearance. Below that point a full circle is impossible from ground level, needing a sighting from high altitude. Getting high enough to see something 10° or more below the horizon is a challenge.

Sunlight passing through the side of a needle in this stable configuration in the sky and appearing out from (that is to say not reflected from) a flat end

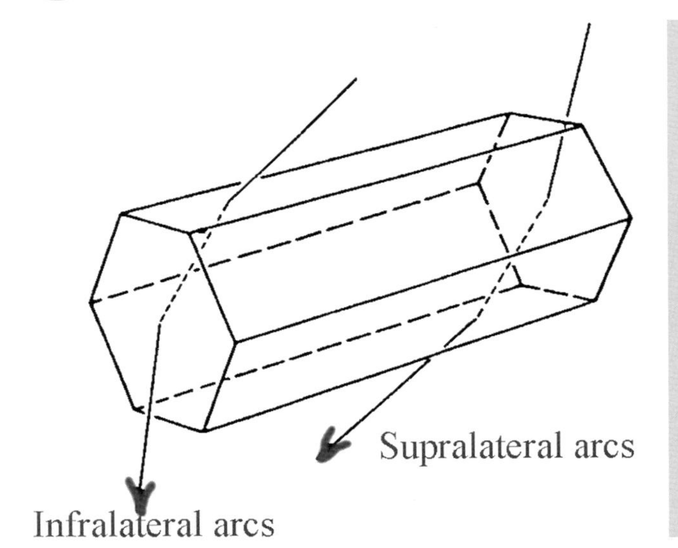

Figure 11.4. *Lateral arcs* Even more stringent conditions are necessary to see these. The sun's height in the sky has a crucial part to play in the manifestation of the shape and extent, brilliance, and so on as light has to pass through the extremely tiny ends of the needle crystals.

in Figure 11.4 gives us some surprising arcs not at the top and bottom of our main one (Figure 11.4), but somewhere in between. These are *supralateral and infralateral arcs*; when the light passes in a similar way through an ice plate crystal, we get *circumzenithal and circumhorizon arcs.*

An extremely rare type of arc, which can span the whole sky, is a *parhelic circle*—that arises from *reflections* off the vertical end faces of the crystals. Needless to say, that's not very bright and is always difficult to spot.

11.5 Bright Arcs Type 2

These arcs are even rarer crystal orientation than Type 1 and extremely difficult to differentiate as their origin is very much the same, if somewhat unexpected. As in Type 1 the basic need is for a very high proportion of the needles to be aligned almost perfectly horizontal and rotating in a vertical axis.

The surprising difference this time is that the needles aren't rotating along their long or needle axis, but enough or a majority don't do this. Then another factor must come into play. The majority of these crystals must be seen with their position such that the flat upper and lower surfaces are parallel with the ground. On a flat surface this would be understandable, but in space not so. Why this stability should exist isn't clear, but it can and does happen.

In these circumstances, with a solar elevation of at least 5°, we came to know of a group of arcs known as *Parry arcs* after the explorer William E. Parry who first described one in 1820 during his search for the Northwest Passage (1819–1820).

They have a very similar appearance to that of Type 1 and are in four similar groups:

- The upper tangent arc, the *Suncave Parry*, appearing with a much fainter arc above it on good occasions. This fainter upper arc appears as though part of a larger outer halo, which of course it can be.
- In between these can appear another arc known as a *Sunvex Parry Arc*.
- We also see an occasional analogue to the Type 1s in *Parry supralaterals and infralaterals*. These also appear further out and as arcs to the 46° halo.
- More interestingly, these refractions duplicate or mimic *the circumzenithal and circumhorizon arcs* normally found with cloud formations containing the plate ice crystals.
- Some observers have seen the *helic arcs*, seen more as a pillar, arising from reflections off the stabilized prism faces.

11.6 Bright Arcs Type 3

In this category we lump everything else not otherwise explicable.

11.7 The Parhelic Circle

Referred to above, this appears from more than a single source. One other possibility exists when the Sun is at a relatively low altitude and light is reflected off a needle surface when that is standing vertically, quite unlike the other orientations so far.

11.8 The St. Petersburg Display of 1790

This phenomenon was seen on 18 June and has never been seen in quite that way since. The basic necessity arises from plates and needles orientated in a slightly different way, with plates still rotating in our two dimensions, but with a flat surface inclined toward us; similarly for the needles.

As one can expect from this complexity, almost anything can and does happen as the Sun rises and sets. The more dramatic affairs occur when the Sun is around a 50° elevation.

The famous display was documented and drawn by Tobias Lowitz, and hence the name (Figures 11.5 and 11.6).

All the halos, arcs, and bows have been explained by the multitude of possible internal reflections, refractions, and external reflections. However, a number of doubts still arise when the brighter arcs can occur by more than one mechanism.

What do you think? Much more important to go out and observe.

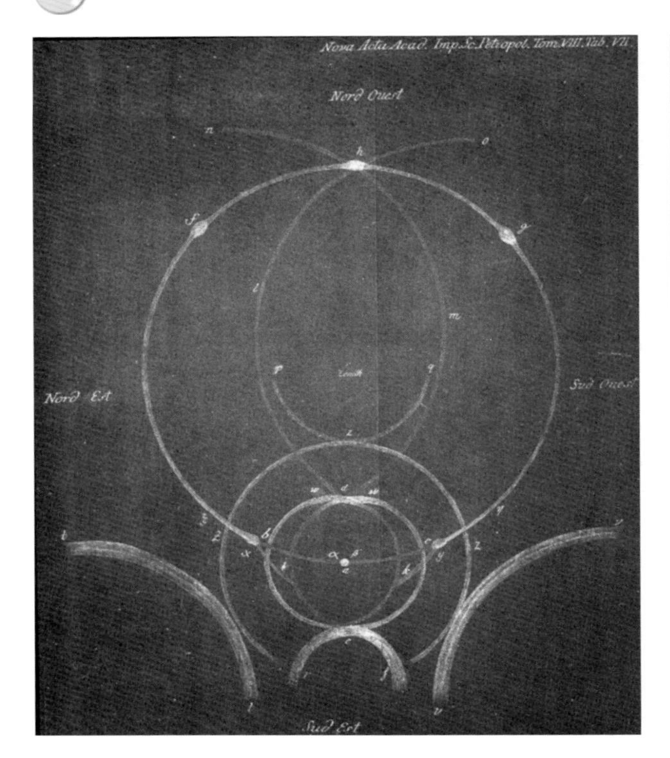

Figure 11.5. *Lowitz picture 1* If you ever see anything resembling these remarkable arcs and halos in the sky as drawn by Lowitz, draw it immediately as he did in 1790 if you forget my instruction to carry a camera at all times.

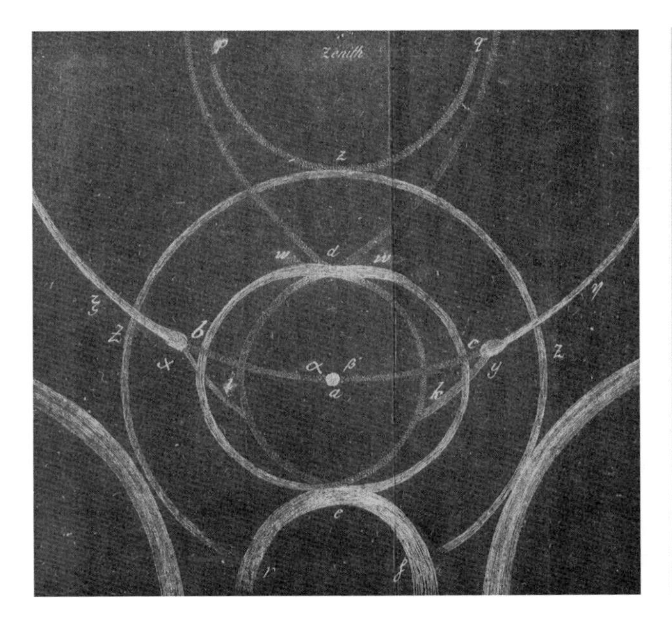

Figure 11.6. *Lowitz picture 2* If you ever see anything resembling these remarkable arcs and halos in the sky as drawn by Lowitz, draw it immediately as he did in 1790 if you forget my instruction to carry a camera at all times.

11.9 What to Observe

As with all our refraction effects, what you see is entirely personal. The person next to you sees his or her own halo and that can be checked in the same way by reference to foreground scenery and/or photography.

The diffuse nature of some halos and their intrinsic low intensity places an absolute premium on your ability to use and be happy with averted vision. All too often, a confirmation of what you see is that much more difficult if the only available witness is clueless on what you're trying to describe and how to see it! Your record might be the only one to be "obvious."

Whatever I've said before on carrying a camera with you at all times is mandatory if you want to capture one of our rarer glows. The basic difference now is that you must become serious and plan a bit more.

The essentials are based on the well-known KISS principle: "keep it simple, stupid."

- Carry a loaded camera at all times, and ready for action. That's where the simple comes in.

Trying a new camera or changing lenses is not an option as many displays only last long enough for you to decide which other lens to find and attach....

- But the one that you carry must be the best you can buy because:

1. It must have a good quality lens, in clean condition. As we're often shooting straight into bright sunlight, flare from bad optics or dirt will destroy every good intention.
2. A zoom does not work for the same reason with ghost images. A prime lens is a lot better.
3. With some halos taking up more than 90°, even a wide angle isn't enough. Be prepared to work with a fish-eye.
4. A bulb or time exposure is also essential if you want to try moon analogues of those seen in daytime.
5. Manual focus option.
6. It must be accompanied by and be operated from a tripod.
7. It should have a plentiful supply of film or memory, as recording the changes over time is by far the most fascinating part.

- An old film camera is ideal for this work as a backup.

1. These cameras come with well-designed lenses and accessories.
2. Modern digital cameras also are good, but the tendency to auto-focus can be a major snag when focusing on diffuse glows.
3. Video cameras with or without still facility perform least well with a limited resolution, and their lenses rarely go wider than a 35-mm equivalent on a 35-mm camera, or about 50° cover. Fortunately, for most arcs and portions of them, this is more than good enough.

- Never miss an opportunity to look up and around whenever the Sun is visible and accompanied by cirrus cloud formations somewhere in the sky. As soon as something is seen, photograph it.
- *Always* have your film processed professionally, and when this is a night scene, ask for the slide film to be returned uncut.
- *Always* keep good notes and accurate timings. Otherwise you just have a memento, not a scientifically useful observation. That means checking the timer on your electronics. I strongly recommend never changing this from UT (Universal Time), or an equivalent wherever you live. Errors introduced for daylight saving cannot be rectified if the solar elevation needs checking too.
- Electronic formats are always a bone of contention as these are improving all the time. Keep the original format from the camera, preferably in RAW, and work on copies. Modern SDs and the like are so small and cheap, and not as big a problem as a single 35-mm slide in a file

Remember that film or memory is cheap, the experience isn't. If all else fails, try to emulate Wilson's brilliant artistry in Figure 11.7 in particular (also see Figures 11.8, 11.9, 11.10, and 11.11).

Finally, never forget all the warnings of never looking directly at the Sun and *never* through optical aid or the camera viewfinder. Whenever in doubt, hold up your hand and use the camera with the other. Even on a pitching boat, this can work. (Figure 11.12)

The big problems in a brief review like this is in deciding what some of these unusual arcs are and what they signify. My experience doesn't seem to be too far from a potential use in weather forecasting.

I've already commented in Chapter 8 that persistent sun dogs often proceed into a period of excellent weather. On the other hand, in most occasions I've seen a good 22° halo that lasted more than an hour, the outcome has often been dire.

Figure 11.7. *Wilson watercolor* Painted by Wilson in 1912 whilst in Antarctica, this watercolor takes a lot of beating for the excitement in glow watching. This picture captures the mood perfectly.

Figure 11.8. *Circumarcdog* What's in a name? This is probably a parhelic circle as it went most of the way round the sky, with this dog at 120° from the sun, August 2006.

Figure 11.9. *Lunar corona and arc* Another "what's in a name?" Photographed in the early 1970s.

Figure 11.10. *Lunar halo with corona* A persistent "hole in the sky" seen for most of a day from before noon till midnight. This picture was taken around 1900 hours. A period of severe weather followed.

My first sighting was at the Montana total solar eclipse on 26 February 1979, Figure 5.2. That bad weather followed me back to the United Kingdom, and we had an exceptionally cold winter with freezing conditions right through to Easter. The next good halo was a year later from The Solent; in the next few days, many people died in that infamous Fastnet Race.

More recently, a lunar halo led to some very nasty weather, and the latest solar one gave us a long period of gales and wet. Here are a few examples I've gathered over the years. The lunar event on Sunday night of 24 March 2002 was so obvious at a full moon that few outside failed to see it (Figure 11.10). The best sightings, when I took my photographs were between 2000 and 2100 hours, although it

Figure 11.11. *Arc seen at NPL* Where better to make an observation of a light phenomenon than at the National Physical Laboratory, United Kingdom? Probably a Parry arc.

Figure 11.12. *Halo and contrail.*

remained visible until around midnight. During the daytime, the halo was also visible around the Sun from around noon, making this at least a 12-h event.

Estimates at the time make the temperature on the ground well below zero and somewhere between −30 and −48 °C in the clouds. The clouds responsible were a mixture of ordinary cirrus and cirrostratus (Figure 4.1).

Our next set of lights in the sky is always associated with joy, or is it if we believe the recent world scene and the Greek's view?

CHAPTER TWELVE

Rainbows

William Wordsworth wrote these immortal words:

My heart leaps when I behold
A rainbow in the sky.
So was it when my life began
So is it now I am a man;
So be it when I shall grow old,
Or let me die

Ever was it so, and shall be forever more, say I as we approach this fascinating subject. For, as you're about to see, there never has been a single rainbow on its own for all to see. What you see is uniquely yours and yours alone. Not only that, but in real time you are carrying that unique sight with you, wherever you go.

Noah rejoiced at that sight we are told, as that signified the end of the Biblical flood. The Greeks took the contrary view, as that was a message from Iris, who portended nasty times to come in war and strife.

Most people have seen two rainbows together; fewer still a third, but in physics lots more are real possibilities, and no less than six are easily calculated.

Nor is a rainbow a phenomenon just of the sunlight. Any really bright light at night can generate the pretty effect, and on top of the list is the Moon. Needless to say, this is extremely rarely seen and then even rarer away from full moon, and hardly anyone's around anyway to make the effort.

I'm considering this as a separate chapter to cover all 24 h, because a rainbow is quite definitely not restricted to most people's impression of being something only seen after a rain shower, whenever the Sun happens to break through.

You can create one to order anytime as a DIY light in the sky as I'll describe later.

12.1 Mechanism

The rainbow everyone recognizes appears like magic whenever the Sun shines from behind onto raindrops falling in front of you. The darker the clouds behind the rain, the more dramatic and striking the effect. But that's not all, the higher you are and the lower the Sun at the time, the higher the bow seems to rise, a very good reason why sunrise and sunset rainbows impress more. A tropical storm at noon can be a miserable affair seen so close to ground level.

Another phenomenon that's only possible from a great height, as in an aircraft, is to see the genuine article, which is a complete circle. I've only seen this once when we emerged from a thunderstorm into a bright clearance full of rain from storm clouds above. What we see on the ground is only that portion of the full circle visible above the horizon—hence the name is correct in being a bow, or bow shaped.

A fine misty rain isn't of much use, and we end up with a fogbow, a relatively miserable affair (Chapter 8).

As with all of our lights in the sky arising from sunlight (or moonlight), the center of the bow we see is precisely at the antisolar (Moon) point in the sky. That point is unique to you and your eyes. Move and the rainbow moves exactly in synchronicity. Your friend next to you will see his or her own personal rainbow. As with the Brocken (Chapter 7) a camera will capture its own rainbow which you can confirm if you have two cameras set up, or if your friend takes a picture at that precise moment. All views seen and compared for parallax will differ from your personal or camera's viewpoint.

For a bow to be seen, we must have perfectly spherical raindrops, which is what we normally get. The heavier the rain storm the better, as each drop is then much larger, well separated from its neighbor (unlike a mist) and in free-fall, more closely forming our perfectly spherical shape. Pear-shaped raindrops are not a normal feature in a rain cloud as water's high surface tension pulls them into spheres in very short order.

As we've seen with fogbows, the drop size is very important. The intensity and how much is seen are all down to those critical matters of drop size and how regular and how closely packed they are. The worse the storm, that is to say not a thin curtain but in-depth, and the higher the cloud base the better from our "viewpoint." No two bows can ever be precisely the same in all respects, particularly when time of day is also a factor.

The next requirement is for a parallel beam of light and that's exactly what we get from the Sun so far away in space. The main bow, which we can also call a *primary* bow, is formed when that light is refracted into the top of the drop, bounces off the inside back surface that acts as an almost perfect mirror, and

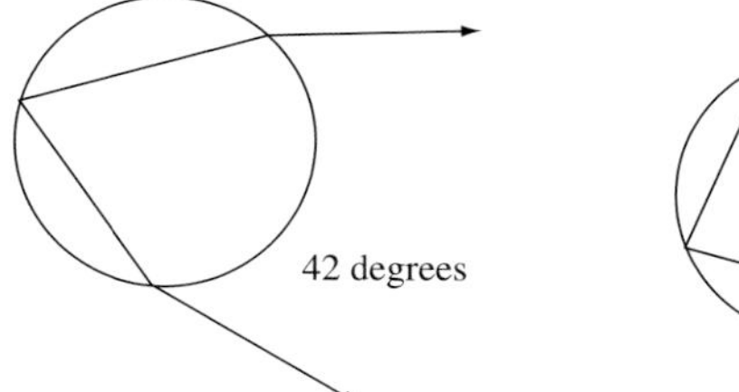

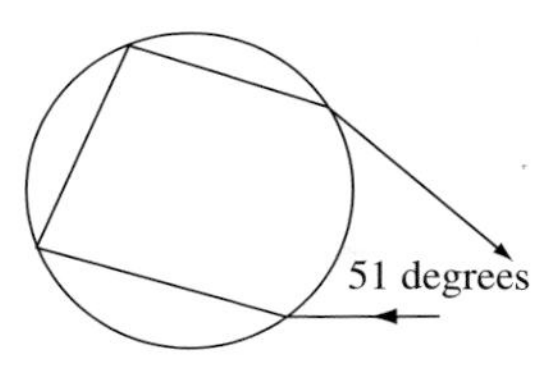

Figure 12.1. Total internal reflections within raindrops lead to Primary 42 degree bow (left) and Secondary 51 degree bow (right).

is then refracted out again at the bottom to appear to our eyes. Light bouncing around the drops loses virtually no intensity as it does so, by a self-explanatory phenomenon known as *total internal reflection*; something we noted with mirage. Figure 12.1 shows how this works.

In practice, the total internal reflections aren't perfect and about half the light is lost in the process. As we see later, the more the number of internal reflections involved (referred to as *orders*) the fainter each becomes.

We get different colors because the raindrops don't fall in a single place, and those slightly displaced from the optimum send back a slightly different color. Well, it's not quite the whole story. The Sun isn't a small spot but a disk, and the angle light entering a drop is ever so slightly different from across that small area. If we take all the effects together, some light is sent back out all over the place and the nett effect is to smear the bow a little.

The important bit of physics remains in that the bulk of the light does come out in a very narrow beam of considerable intensity. The name for that very narrow light beam is a *caustic*, which is an extremely small-angled beam compared with all the other possibilities that spray out in all other directions. That is a partial explanation for the diffuse nature of fogbows, when that effect then adds to diffraction, and it all becomes part of the background blurring.

For simplicity, I've not shown all these other complications. These complications include what are known as *supernumerary bows* where the light from these slightly different angles causes interference and lead to a whole series of much fainter bows. In normal daylight these aren't seen, or just suspected at the limits of perception immediately inside the main bow.

However, under section "What to Observe," it's fun to see whether or not you can prove me wrong in that these illusive extras are so rare. I've seen two of them together on good occasions but don't have decent photos. Figures 12.2 and 12.3 show rare occasions when part of this phenomenon showed up.

My best description is greenish to blue as I can see quite well in those colors and suspect that a whitish overall is probably a nearer or better description for many readers.

It was way back in 1600s that Newton first studied light and how the spectrum arose in a scientific fashion. One tool he knew about was a glass bowl of water, a larger version of our simple raindrop. He then validated his new calculus in mathematics to prove the angles at which a rainbow appears. For our primary rainbow this is 42°.

Figure 12.2. *Supernumerary Rainbow* Check the left-hand portion. A "grab shot" might turn out to be scientifically useful, Ushuaia 1998.

Figure 12.3. *Rainbow portion* A telephoto lens can work better than wide-angle lens to bring out the supernumerary in this isolated light in the sky.

The way the light is refracted through a water droplet and redirected back to you depends on a number of factors. These include the refractive index of water. Fortunately, for rainwater, which is about as pure as it can be, that's close enough to a constant 1.33 for red light and a bit higher for blue at 1.35.

The higher the refractive index, the more the blue end of the spectrum is bent away from the original beam's path, red being much less affected. It's not generally appreciated that a substance doesn't have a single refractive index although that's often "quoted." What they mean is the measured figure at a specified color, often green or sodium light.

Each wavelength of light has its own Index that seems to contradict Einstein's famous Law here. It doesn't because the ultimate limit is in a vacuum, as we saw with Czerenkov radiation. Refractive index is best defined as the relative speed of light in the two media; which in our case is plain water and air.

The same process works with gemstones. Diamond has a very high refractive index and the way blue is dispersed gives all the "brilliance" that we covet so much.

Ignoring the mathematical formulae, we can calculate that red light comes back at us from an apparent 42.6°, slightly more than for blue at 39.7°. The real rainbow angle is always quoted as 42°, but as you can see that's not strictly the average value, particularly for those seeing well into the violet. The full angular width of the bright rainbow we always see works out at a little below 3° under perfect conditions.

Under those same excellent conditions, we always see a secondary (*second order*) rainbow outside the main one. This is some 10° further out and only 43% as bright as the main one, with a reversed color sequence. This bow is formed when the raindrops are big and heavy enough to reflect enough light to be seen from two reflections from within the back surface. Small drops or light drizzle rarely exhibit a secondary (Figure 12.1).

In this secondary bow situation, light enters the other way round at the bottom of the drop and re-emerges at the top, and our second of the two reflections (as in any mirror) reverses the image. In this case, the image we see is the same as before for the main bow and is of the Sun at a particular wavelength of light.

Doing the calculations again we find the expected lower deviation for red around 50° and for blue another 2.2° further out. An average figure of this is always quoted as 51°. This secondary bow is always narrower, and is fainter as you might expect from the lower light levels due to the losses on the way noted for the primary bow The physical laws of refraction (+ reflection) are what count here, not a contrast effect or our eye's sensitivity.

As with all refraction, red light deviates least and blue/violet most. Keep that in mind every time you see a spectrum portrayed (Figure 1.1)! Drawings and paintings for the two rainbows must have the two reds together, as it were, on the inside.

Light at these bows can appear as though it's channeled from the space between them to appear as the bright bands we see. In fact the light isn't like that but sent off into a space where you can't see it. The effect is always the more awe inspiring when you can see the land behind, that is to say below the main bow as in Figures 12.2 and 12.4. The region between the two bows is known as Alexander's dark (not ragtime) band.

Figure 12.4. *Rainbow* All the main characteristics show well in this picture, particularly Alexander's dark band.

When we come to the third (*order*) rainbow that's from three internal reflections within a drop, this would appear to be impossible for a very simple reason. It's not where you expect it to be but behind you, that is to say as a ring around the Sun. Our third rainbow's calculations indicate something much the same size as the primary, or 42° across.

Then there is the problem of seeing such a faint ring against the full solar glare. A huge amount of light is sprayed out into a disk or aureole that more or less fills the space taken up by our third-order affair, which has dropped in intensity to about a quarter that of the primary.

With so many internal reflections from a light source of appreciable size, some spreading of the bow's width also occurs to drop the apparent intensity further below our theoretical quarter.

The clincher will have to come when someone sees all the colors correctly displayed round the Sun behind you in a clear sky at the same time as a main (with or without a secondary) seen in front. The chances of rain falling that neatly in two separate places and of the right type make this a highly rare thing.

That this theoretical third bow is well known is amply confirmed from a seventeenth-century contemporary of Newton, none other than Sir Edmond Halley. His calculations placed the bow firmly round the Sun, not where we see ours.

The same applies for light subjected to four internal reflections where the amount of light available to see has dropped right down to about a sixth that of the primary, and of a comparable size, and more diffuse overall. These are the "mirror images" (pun not intended) of the one and two reflections forming the ordinary bows sent back to your eye.

Some have claimed to see the third one, but the descriptions seem to match a halo, or corona, or similar colored lights within the bright aureole. As some storms arise immediately after the passage of some cirrus or similar, this thinking is the most likely explanation of these considerably smaller rings, arcs, and bows. Think in terms of those sightings generated by refraction from ice crystals a lot higher up.

Try my suggestion below with a strong light behind a water fountain and see if your thinking is the same.

The five internal reflections' bow also cannot be seen because that's somewhere on the ground in front of you. In theory the bow from six internal reflections ought to be seen somewhere in the vicinity of the two main ones. With that number of internal reflections the absorption from the longer light path and small losses at each reflection all add up so that emerging light will be a tiny fraction of the main events and will be swamped. Modern laser lighting might show something.

12.2 What to Observe

Rainbows are something we all enjoy, so do I need to say more?

Because each time we see a bow it's a unique and uplifting experience, after all the dismal rain and stormy weather preceding it, the temptation remains to photograph what we see. I suggest that's a very good reason for doing it, in just those words "a very good reason."

As I started to research this chapter, I delved into my film files and found many from my past history that I'd forgotten, and very pretty they still look today. What was much more fascinating is the way those old pictures showed up some of the features I'd been blissfully unaware of until then. The supernumerary bows and other features that now made sense, for instance.

Always carry a camera as you never know how (scientifically) interesting a picture can be to future historians going through your effects, long after you become a sky glow.

Whenever you see a good sample, photographic rules couldn't be simpler. Set the camera for a normal daytime shot, adjust the lens to infinity (very important with autofocus) and shoot away, preferably on a tripod. For some reason hand-held shots always seem to disappoint, and that might be due to camera shake from the excitement of the moment.

For 35 mm, a standard 50 mm focal length will capture enough of one end of a bow to do the necessary, and a 28 mm or wider-angled lens should capture the lot, if somewhat poorer resolution at the edges. As with all lights in the sky, a slow-speed film will capture the colors and contrast a lot better against a darkened sky than a fancy high-speed version. All modern digital and video cameras will give you a good result. The picture's perfection then lies in lens quality and its cleanliness. Dust and grime must be checked and removed as we are after the transient and fainter parts on view, not the brighter parts that can be seen at any familiar apparition.

At night, the rules are much more stringent. The first imperative is to check as a rainstorm ends at any time around full moon. As in daytime, a tripod is a must and spotlessly clean optics. After that, the same rules apply as in the day. Shoot off as much film or memory as you can. I show no moonlit rainbows, as I have none.

That is your first challenge to film something few will ever see. These are a few other real challenges to recap the basic mechanisms. Do you see any of this?

- The bright arch under the main bow.
- Alexander's dark band.
- The dark sky outside the secondary.

That this shows as a slightly "brighter" sky density in that much larger outer backdrop.

- Any supernumeraries.

If so, how many?

What color do you see in them? Any color seems possible, part of the science yet to be recorded.

- As a test of eyesight, can you see a strong violet on the outer fringe of the secondary bow, or just a pale blue smudge.

Check that immediately against what you see in the main bow, where supernumeraries can and do appear white and distort the appearance in that innermost overlap region(s).

- Does the thickness vary as the rainstorm moves, and does that change rapidly?

It normally does. The apparent angle seen remains absolutely constant, but the rapid changes in apparent size all boil down to a distance effect as the main rainy band travels relative to you and the Sun. Something that's very similar to how we see a glory in cloudbanks and our bit of DIY below.

A rainbow can be seen under extremely clear air conditions up to 60 miles/100 km away, right up to a few yards/meters.

- Sometimes a mist follows the main event. If the Sun continues to shine on that, try to capture a fogbow and demonstrate in real time the fundamental differences between reflections and interference effects from diffraction.

Even from this briefest of checklists, it's obvious that rainbows aren't as simple as many think. The plot does thicken somewhat in the denser and more turbulent air around dawn and dusk.

At these times, when you're out to view the other glows, never give up in disgust at the weather and thereby miss the opportunity to do some science whenever a rainstorm passes. The geometry is almost perfect to see a bow against a dark sky and you don't have to look up to see its full extent, as the bow at the extremities will not just skirt the horizon as at noon, rather stand up. Often those pieces are all you see due to perspective and the clouds almost merging with the horizon.

One of the more fascinating features to come out of rainbows seen at these extremes of the day has to be the way the atmospheric density and humidity alters the primary colors.

The closer you see a bow with the Sun on the horizon, the more extreme this factor becomes. Never miss out viewing at those times to see if any of these are true:

- A near or even totally red rainbow. That's right across where you'd expect to see the spectrum. A secondary bow is extremely rarely possible. Is that single the main one after all?
- A rainbow seen in rain as the Sun has set below the horizon behind you into a clear sky, where reddening of that light isn't a factor.

  Very much as a reddening of sunset at these times we can explain the red bow in the same way, but one from a set Sun? Mirage effects apply here, sometimes, or the sunlight is directed at our antisolar point, which can be some many miles up in the sky. Rainbows up to 20 min after sunset have been reported.

It would appear to follow from these mechanisms and suggestions that you can only see a single set of bows. If the conditions are right, that's quite wrong.

Many people see more than a single lot, with the bows intersecting at some very odd and strange angles. The explanations can be bizarre and unique to each occasion whenever the main light source (normally our Sun) is reflected off something and creates a new light source. This new source has its own antisolar point for us, the observer, and that creates a new set of bows.

The classic situation arises when you see a rainbow from across a lake and that reflects upward. Even more interesting is a bow seen from sunlight reflected off a lake behind you, mingling with its almost equal intensity from the "real" sun.

The bow we never see from multiple internal reflections from within the drops can appear, again as if by magic, from these normally forbidden directions.

The weirdest example of this rainbow appearing in the wrong place occurred in summer 2006. While traveling by train in London, a rainstorm had almost disappeared over the horizon, forming a black screen over Canary Wharf. The setting Sun reflected off the building's glass formed a perfect rainbow a few building's distance away, but just at that size and height—which leads us to our final piece of fun, making a rainbow in order to show and study at leisure all these weird possibilities.

At any time water droplets appear and are illuminated by sunlight (or any bright spotlight behind you), expect to see a rainbow (Figure 12.5).

The perfect way to generate a complete circle is normally totally ignored when the kids are having fun round a pond, or playing with a hose. When vigorous pressure is applied in the hose bigger water drops are formed which act as rain. Whenever the whole crowd is subjected to this torture, and the hose held high, a complete rainbow should be seen. A friendly fire brigade is often a good place to start next time they are practicing or training.

The higher the ladder is the better, and the water curtain will fulfill our criteria of a high backdrop of uniform water droplets, illuminated in normal bright sunlight. Capture the event on film and check the angles and colors seen.

Our next step is to check what happens when the light is coming from behind the water droplets. This simulates all the cloud effects in previous chapters and the main and essential difference is the ability to see rainbows formed by the normally forbidden three and four internal reflections. Fountains and spray patterns over ponds are a good place to start.

Although this isn't something to be attempted at home, almost any other liquid in (largish) droplet size will show rainbows. Oils and any other transparent

Figure 12.5. *Niagara, 1970* Always keep old pictures. This shows how a rainbow can form anywhere water droplets are sunlit, as here below the horizon.

liquid will do. The interesting feature of these other liquids arises from the quite different angular size you see, not the 42° with water.

Any time an accident happens and some liquid is sprayed about, don't stay as a morbid onlooker, but go for some serious science. The bow angles can be used to calculate the refractive index of the liquid and confirm the substance to a subsequent accident enquiry! Check the color sequences, of course.

Perhaps the safest way of seeing rainbows at nonrainy times is at a waterfall or spray from a boat. Anywhere, in fact, at anytime this unexpected light in the sky can appear (Figure 12.5). Sometimes these pictures, as here, have historic value in their landscape changes decades later.

A final time to keep a watchful eye open for a rainbow is anytime there's a heavy frost. Sometimes rain falling from a relatively warm cloud will freeze into droplets, or be frozen that way before a normal snowflake can materialize. Though it is extremely rare, this has happened. This isn't necessarily an example of Diamond Dust, but it might be.

Much more commonly, water freezes into one or two crystal shapes and when that happens and conditions are just right, we are set up for the halos and arcs of lights in the sky (Chapter 11).

Rainbows normally follow or precede our next dramatic lights to be seen at any time. The only extreme lights that can harm us directly is lightning.

CHAPTER THIRTEEN

Flashes

13.1 Lightning

"Flash, Bang wallop," followed by "what a picture" is a well-known introduction to a song, but it could also be the introduction to a modern version of a well-known UK nursery rhyme that ends "All fall down." In this case it's nothing to do with the plague, just telling us something much more immediate and deadly—lightning.

We've all heard tales of people killed in a storm when standing under a tree, or being too close to something metallic. The problem is that lightning can be just as lethal anywhere and strike in the open air, totally without warning. A classic case happened at a football match in October 1998 in what we knew as the Congo. Lightning struck the open pitch and no less than 11 or half the players on the field were killed outright, and a further 30 others in the crowd got very severe burns. Other similar incidents at football matches have occurred over the years, and on 4 May 2006, nobody in Dumfries, just outside Edinburgh in Scotland, seemed to remember this dire warning from the past.

The spectators and players blithely ignored an impending storm. A lightning bolt knocked several players to the ground, and quite amazingly the spectators raised umbrellas whilst running for cover as the rain started. Several got severe electrical shocks, but nobody was killed on that occasion.

The year 2006 seems to have been a bumper year for lightning strikes and near misses. Bolts have jolted children out of bed when lightning passed through buildings, and these were sometimes accompanied by ball lightning.

These casualties over the millennia assuredly make lightning by a long chalk our most hazardous light in the sky.

13.1.1 Mechanism

How all thunderstorms are created remains quite unclear even after centuries of study. The basic mechanism lies in an electrical charge built up on ice and water droplets, which then get collected together by movement and air currents within cloud structures.

An electric or static charge built up on water droplets is easily dispersed when they collide, one would think, but some clouds still generate storms under those relatively warm conditions. The trouble really gets under way when the water freezes rapidly as it is conveyed to greater heights. In storm and tornado conditions the ice crystals crash into each other so vigorously that they're broken up and then recombined in endless permutations.

The colder the ice is, the better insulator it is, and static charge is not reduced but builds up into the huge pockets that eventually release all their energy as lightning. During this process of fragmentation and recombination the minute ice particles can be fused together into much larger conglomerates than simple rain drops, into our familiar hailstones.

These hailstones can reach gigantic proportions, and many recent storms in America have led to as much damage from these as from the main storm winds and lightning. Hailstones large enough to shatter windscreens, for instance, can also kill and maim.

We all know how vigorous the winds within clouds can be that create all that havoc when ice particles are repeatedly crashing and exchanging charge. Those internal winds are very obvious when we see storm clouds approaching and totally changing shape in a matter of minutes.

The dangerous winds created are the main reason why aircraft avoid electrical storms. The up and down drafts can break up a plane structure, and the electrical effects are (actually) the lesser worry.

When the voltage difference between the charged regions reaches a critical level, the charge is then transferred as the air insulation breaks down, creating a spark on a gigantic scale.

The voltages concerned run into millions of volts. The energy is also huge as we'll see for "sprites."

Most of the lightning flashes we see as lights in the sky are a simple exchange, neutralization of charge is a better description, within those pockets of charge built up by the internal air movements. The lights we see simply light up the sky, with a rumbling background of thunder, and cause no damage at ground level. Crackles and hisses also cause interference on radio reception, and that is still used to monitor thunderstorms worldwide.

The mechanism also seems to apply in other atmospheres, and lightning has been observed on other planets, containing totally different gases. Lightning seems to be something inherent in any atmosphere in motion, where large temperature gradients occur at the same time. Some of the worst thunderstorms seem to occur when the temperature gradient is large, as in cyclones in the Atlantic and typhoons in the Pacific.

Until recently it wasn't appreciated that there are two types of electrical storm leading to lightning, and I'll go into more of this important difference in "sprites."

Meanwhile, we cannot leave this historical review without bringing to the fore one of the more enterprising experiments of all time. That must be Benjamin Franklin's.

Taking time out from his political expertise in the American nation's beginnings, he flew kites to find out as much as he could of the electrical forces working in the clouds. Today's politicians of a lesser stature and totally lacking scientific curiosity would be much more concerned to stop such basic research in the name of "Health & Safety."

13.1.2 What to Observe

What's ordinary about lightning? Nothing at all as each storm has its wonderful individual and unique characteristics and frightening aspects.

13.1.3 Simple Enjoyment

By far the best option is to view from a distance and certainly within a protected structure. All modern buildings are designed to withstand the most powerful lightning strike, particularly when of steel reinforcement construction.

There are two other options when caught out.

1. If entirely in the open, forget your fancy clothes and lie down. Lightning strikes the highest point, your head when standing up. There is the classic case in America of a man who was struck by lightning on several occasions when in a boat on a lake. After about six "warnings" the next one killed him.
2. If in a car, stay in it as that's the closest thing to a Faraday cage and lightning and electrical charge stay on the outside.

However, if you've opted for a modern vehicle made of fiberglass or carbon fibers, then rule 1 applies. The engine is made of metal, to say nothing of the seat belt buckles.

13.1.4 Photography

Taking pictures during daylight hours is always said to be the ultimate test. Call it luck. A lightning flash is over in an instant, and nobody's reaction time is ever going to cope with that. Then there is the added complication with digital of the delay time after activating the shutter (Figures 13.1).

By far the best option these days has to be a video camera, which is set in motion and allowed to run on a tripod or a similar steady mount. Camera shake due to the excitement of the moment is too probable when handheld. The timed sequence can sometimes be of scientific value, but is so much easier to grab a nice picture as a frame.

(C)

Figure 13.1. *Lightning* Photographing lightning in the day is notoriously difficult. This sequence in the 1970s relied on no modern electronic gizmos. Old cameras still work!

Sometimes, extremely rarely, the strikes occur at more or less regular intervals, and that's about the only time daylight photography is ever going to have any guarantee of success.

At all other times and particularly at night, photography with a conventional or straight digital camera is greatly simplified.

Set a camera on a tripod, stop down the aperture to the maximum extent and open the shutter and leave it until a flash is seen, then close the shutter. The light from the flash will achieve what a normal flashgun will, but on a magnificent scale. Then keep going until you run out of film or memory. Leaving the shutter open for more than a single flash is cheating a little, but it does economize on film and can often lead to some "striking" pictures of a composite of several flashes to treasure.

The choice of backdrop is entirely a matter of personal preference; lightning striking over a town or city always seems to be best appreciated, although the reason for that psychology is not obvious. Lens choice is then important and a wide-angle yields best under those circumstances, whereas a tele around the equivalent of more than 85-mm on a 35-mm camera works better for open-air shots. The intricate detail in the strike shows clearer the more the magnification employed.

Whenever the storm is sluggish and flashes infrequent, a useful tip is not to close the shutter, but to cover it with a lens cap, or even use your cap, removing when things hot up again. The important factor remaining is to remember to close the shutter when packing up, otherwise your masterpiece (always the last?) will be fogged out.

Lightning photography is another occasion when an old camera with its standard lens comes into its own. Getting it wet or muddy is no great loss. Similarly, out-of-date film is quite usable as we are seeking a very bright image on a dark background. Films of 100 ISO or slower will do the business extremely well (Figure 13.2).

Our next two forms of lightning are much more difficult to photograph. One from the probable terror of the occasion and the other from the great heights involved.

Figure 13.2. *La Palma volcanic storm* A very rare light in the sky, a lightning display due to volcanic activity, photographed in 1975, long before the observatories were built there.

13.2 Ball Lightning

Even more mysterious than our normal thunderstorm, perhaps this is our most perplexing light in the sky phenomenon. It can also be the most threatening, as it appears quite unexpectedly and seems to bear no relationship to anything we experience in a normal day.

Something definitely not to be tried at home, even when it comes to visit you.

13.2.1 Mechanism

Ball lightning can occur at any time of day, and there are plenty of records that a ball lightning strike doesn't always occur with a major storm.

A classic sighting seems to follow a pattern with the following features:

- A large rotating ball of (mainly) water vapor.
- That ball has highly charged water vapor at the periphery.
- The ball is moving, but not necessarily rapidly.
- A high proportion seen, in some statistics up to half, occur indoors.
- People experiencing them are often up-close and personal.

Reports vary on the size, and these can be fist-sized up to huge. Sightings in the open air have been claimed to be UFOs as the brightness reported can be like full moon or a 100-W light bulb.

Even more puzzling are reports of ball lightning inside a Faraday cage, such as an aircraft, and some witnesses actually handling them.

Totally unconfirmed reports of ball lightning effects being generated within a microwave oven leave much to be desired, and border on the fanciful aspects of UFO. However, the reports of sightings close to gas works and similar changes to the atmospheric composition do require more study. Analogies of the visual and photographic similarities to soap bubbles make much more sense in that context.

Whatever the mechanism, it's clear that no single theory can explain all sightings and extreme caution is advised whenever seen.

Studying ball lightning is one topic we can safely leave to the experts and not one to try at home when something comes to visit. As in a normal thunderstorm, "Ball lightning Chasing" is an ultimate folly, but then, if you do get a visitation.

13.2.2 What to Observe

The statistics on these peculiar lights in the sky indicate that they're nowhere near as rare as once thought. An estimate of up to a thousand an hour worldwide is now accepted. Of these, about half are seen in a domestic environment, that is to say, indoors.

Whilst staying well away try to judge all these features:

- *The shape*: The definition of a ball seems to fit most reports. American or Australian football shapes could confirm a "soap bubble" hypothesis affected by air currents.
- How the shape varies with time?
- *How long do they last*: A normal lightning flash is, effectively, instantaneous. These can last for several seconds, sometimes much longer the larger they are.
- *Size*: Some reports record something easily fitting into a fist, others outside talk of a football pitch across. Most reports talk of a football size.
- *Speed of movement*: Changes in direction are frequently reported to complicate this judgment. Human beings are notoriously fickle at speed estimates.
- *Brightness*: This can also change with time.
- *Color*: This aspect is also another fickle human failing in terms of accuracy.

If you can achieve any or all of these, you're doing better than I would. At least there are no reports of radioactivity, which rules out that avenue of doubt.

It is imperative not to get too close and handle as some have claimed to do in the past. Many examples of ball lightning end in an explosive display, often with structural damage to property or extremely nasty burns to flesh. Some of these burns seem to occur from metallic jewelry reaching red heat due to electrical induction.

That is our final explanation for the source when a sample of ball lightning has been reported, it's nothing natural in the conventional sense, rather our lifestyle.

Sometimes the electrical currents within our modern domestic fixtures, particularly those using 110 V, can be excessive. The magnetic induction when the circuit breaks or fuses can lead to some pretty powerful magnetic fields associated with those currents.

Next time you blow a fuse, don't be surprised to see a mini-ball lightning appear somewhere close by. Is it just a familiar smoke ring or another example of ball lightning?

Definitely don't try to make this one at home, as insurance cover is not an option. Much better that we pass swiftly to the really gigantic balls and glows in the sky, our final study.

13.3 Sprites

Until very recently, airline pilots' reports of seeing bright lights in the sky well above the normal weather heights were dismissed as fantasy, or worse comments on their career prospects. That curt dismissal persisted in spite of numerous anecdotal sightings for more than a century.

Then these lights in the sky were seen from space as hugely brightly colored streamers going way out and visible up to over 40 miles (70 km) higher than the ordinary storm clouds below.

A totally new field of research was born.

13.3.1 Mechanism

That something unusual was going on became obvious when some very short-duration gamma ray bursts were identified as coming from within the atmosphere. The duration of approximately a millisecond ruled out nuclear tests as the Compton Gamma Ray Observatory saw only energies in the 1+ MeV domain.

These gamma ray bursts mustn't be confused with the gamma ray bursters discovered since those days, which are the biggest explosions known but coming from the depths of space. The duration and locale were clearly terrestrial. The gamma ray bursts aren't sprites but seem associated in some way, together with huge radio frequency radiations, at the other end of the spectrum.

A chance sighting in 1989 as a film image led to a whole series of positive sightings from the Space Shuttle in the early 1990s, and the new science started.

- Sightings from aircraft showed that the lights were of a blue color, usually in narrow conical structures rising directly out of the storm clouds causing a lightning display on the ground. Velocity upward clocked in around 65 miles (100 km)/s.
- Associated with these blue jets are massive red regions above active storm clouds. These vary in size from weakly luminous spots and extensions to the clouds to huge areas reaching up to around 53 miles (85 km). The brightest regions occur in altitudes of 40–46 miles (65–75 km) and some associated blue tendrils underneath go down to 25 miles (40 km).
- A characteristic of these red sprites is the enormous extent, sometimes more than 30 miles (50 km) across, and taking up tens of thousand cubic miles or kilometers. That puts them on a par with aurora for sheer size and extent.
- Best estimates of these early sightings were staggering as the optical energy seen amounted to some 10–50 kJ, or 5–25 MW, the power output from a decent power station. However, that was only part of the story as the optical energy equates to a mere 0.1% of true. That means these early sightings amounted to an optical power of 10–100 MJ, that is an unbelievable 5–50 GW.
- Those early sightings also confirmed that discharges occurred in multiples of two to three each time. The blue jets and red areas became known collectively as sprites as their regularity became accepted.

The basic mechanism of sprites is associated with a normal lightning cloud or storm.

Here a cloud containing negative charge is our regular feature, and then we see a strike hitting the ground at any high point that can conduct or assist electrical passage. That's not entirely correct, of course because extremely high speed photography shows that many or most lightning flashes actually travel upward to discharge the cloud that way.

These normal, if anything on that scale can be regarded as normal, clouds are responsible for around 90 percent of all thunderstorms.

Sprites are more particular.

If the cloud's convection is such that the charge built up ready to discharge is positive, those are the strikes that tend to preferentially hit buildings. These are by far the more dangerous option, and these also are the main culprits in starting forest and other vegetation fires. The scene is then set for a possible sprite.

- The currents involved with the remaining 10 percent of storms are genuinely massive. Instead of currents to ground of (only!) 20,000 A, these clock in at five times that or up to 100,000 A. Hardly surprising, then, why those rarer storms cause so many fires and structural damage.
- Part of the explanation and mechanism involved arises from the flash originating in the storm's anvil cloud structure. That is to say, the highest portion, which confirms the higher energy needed to reach the ground from higher up.
- Sprites are the associated counterstrike for a ground one we see, and the atmosphere is set glowing way above, very much as with an aurora and of comparable intensity. Their normal color in the glowing upper atmosphere is a bright red and blue below.
- The blue jets or spikes are not associated with nor aligned with the earth's magnetic field. They truly do their own thing.

13.3.2 What to Observe

A study of these sprites is continuing but any normal observer is unlikely to see one from ground level. The best ground-based viewpoints to try are in Colorado, a noted thunderstorm hot spot in high summer.

Visual sightings elsewhere remain elusive from the ground, as the intervening atmosphere must be both clear and dark enough. Movement is upward from the top of our storm clouds, which in turn can be as much as 12 miles (20 km) high. Camera shots have shown sprites appearing way above the normal thunderstorm ceiling appearing in a genuine flash, or instant. The eye cannot always respond fast enough.

Modern digital and video cameras with CCD or similar sensors perform much better than conventional film as the main color seen is red to infrared. That's another reason why visual sightings often fail due to color blindness.

As we saw in an ordinary thunderstorm, we're extremely unlikely to see anything during the day, and even have less chance of photographing something. Night observations show up the coloration properly.

These guidelines increase your chances:

- Wait until a storm is some 120–200 miles (200–300 km) away.
- Wait until total darkness, with a very clear atmosphere showing stars *above* the storm clouds. That is a rare occurrence, requiring the storms to be that long way off with no intervening cloud, atmospheric muck, or light pollution. Anywhere with light pollution will always be a forlorn hope, and an explanation why rarely seen in the past, because our next vital requirement is-
- You must be totally dark-adapted. If you cannot see the Milky Way think of something else to do.

The overall brightness is no more than a regular aurora.

- At those distances you will only see at the limits of your detection, which means operating at the crossover threshold for your eye's rods and cones.

If you're not used to working with averted vision (Chapter 5), this type of work is next to impossible. These extremely transient events will, literally, appear out of the corner of your eye, and looking straight at the storm simply won't work.

- Don't be distracted by ordinary lightning flashes and continue with averted vision. It may help to shield the eye from the regular stuff as night vision can be destroyed for the critical interval when a sprite occurs.
- What you're looking for is a brief flash of light appearing above the clouds. There'll be no doubt as the flash will be huge even from that distance. Expect an angular distance of 10–20°, which corresponds to three to four times the Big Dipper's pointers.
- If you're blessed with good night and averted vision, make some estimate of the structure.
- Don't give up. Expect a score of less than a 1percent sighting compared with the other flashes going on. The confirmation that you've actually seen something will come from the pictures taken at the same time.
- Try to relate your observations with radio turned on at the time. Then check with the other records via the web and the like.

13.3.3 Sprites and Some Predictions

Sprites seem to generate both red and blue jets, although not always seen visually together. It is also clear that the huge energies involved do alter the atmospheric chemistry at those rarefied levels. This is the height where we expect to see some of the other glows that we've met on the way, particularly those associated with the ozone layer.

The worrying aspect of sprites comes from the huge energies involved and the regularity, probably increasing. One of the many chemicals produced in thunderstorms is good news when this occurs in the weather zone, as that's a major source of nitrogen oxides, which then descend in rain and create nitrogenous fertilizers; the main source, in fact that's not made in legumes and other plant roots.

Sprites operating at these higher strata dump precisely those nitrogen oxides that we don't want because of their catalytic ozone-destroying properties.

Observations and deeper studies of sprites are destined to become a major feature in global warming hypotheses. Another classic chicken-and-egg situation: "Are sprites derived from some external influence on the climate, or is the overall warming leading to more energetic storms." Time will tell.

What is abundantly obvious today is that sprites are very much associated with currents within the earth, and that's the reason why the chicken-and-egg analogy applies. Do the electrical storms built up by air currents in the cloud influence the ground currents, or is some earth current happening way below ground level going to induce the charges within the clouds?

As we noted for the similar aurora glow, the earth's magnetic field is dropping at an unprecedented rate over the last 150 years and solar activity is on the rise. Ultraviolet and other storms from the sun are also on the increase and then interact at these very layers in the atmosphere.

It's fitting that our most energetic light in the sky should turn out to be the most enigmatic and potentially most crucial in the weather we experience below.

Climate change, even a sudden tipping point as has been the normal in the past, is on the way, and with it the type(s) and frequencies of the lights in the sky.

One of the unfortunate aspects of all these flashes in the sky is the regularity with which computers and power networks get zapped.

Happy viewing whatever takes place after I shut down the computer.

CHAPTER FOURTEEN

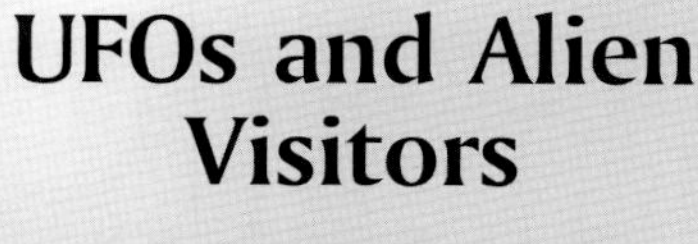

UFOs and Alien Visitors

John Watson

> I would more easily believe that two Yankee professors would lie than that stones would fall from heaven.
>
> *Thomas Jefferson, 1801*

In 1947, an alien spacecraft crashed in New Mexico near Roswell, and the dead bodies of the humanoid alien crew were recovered and subsequently autopsied. The American government initiated a massive cover-up, which has never been properly exposed.

Well, it's possible.

To me, it doesn't actually seem very *likely*, although it is hard to prove that it didn't happen. It is always very difficult and often impossible to prove a negative. It just *might* be a cover-up.

One evening a few years ago, a friend attracted my attention to a brilliant "UFO"—by which he implied "maybe an alien spacecraft", despite the fact that the term "unidentified flying object" is unspecific—low in the southeastern sky. Jupiter, I thought. Maybe Saturn. But the light, although about the right brightness for Jupiter and in roughly the right part of the sky, was too white and turned out to be moving slowly and steadily eastward. I guessed that it was a helicopter with a searchlight. We are a few miles from an air force base where Chinooks are stationed. Nighttime exercises aren't uncommon. The brightness of the light and the fact that it was maintaining about the same apparent height above the horizon implied that it was probably an aircraft of some sort. There is no way I can prove that it wasn't an alien mother-ship scanning for suitable abduction candidates, but I believe the balance of probability is against it.

Let's begin by taking a look at the evidence for intelligent life in the universe. This is important, because evidence of intelligent life elsewhere would give strong circumstantial support to the "UFOs as aliens" hypothesis.

14.1 Are There Any Aliens? The Drake Equation

The first conference dedicated to the Search for Extraterrestrial Intelligence (SETI) was set up by Frank Drake in 1961, who worked at the National Radio Astronomy Observatory in Green Bank, West Virginia, and J. Peter Pearman, who was on the Space Science Board of the National Academy of Sciences.

While getting ready for this conference, Drake worked out what has become famous as "the Drake Equation." The idea of the equation is that if you can put in sensible estimated values for each of the terms, it will give you an approximation of how many intelligent extraterrestrial races there are that ought to be willing and able to communicate.

$$N = R_* \times f_p \times n_e \times f_l \times f_i \times f_c \times L$$

The terms mean the following:

N the number of civilizations in our galaxy that we could expect to be able to communicate with at any given time
R_* the rate of star formation in our galaxy
f_p the fraction of those stars which have planets
n_e the average number of planets which can potentially support life, per star that has planets
f_l the fraction of the above which actually go on to develop life
f_i the fraction of the above which actually go on to develop intelligent life
f_c the fraction of the above which are willing and able to communicate
L the expected lifetime of such a civilization

Drake used the following values:

$R_* = 10$ per year
$f_p = 0.5$ (e.g., half of them)
$n_e = 2$
$f_l = 1$ (e.g., all of them)
$f_i = 0.01$
$f_c = 0.01$
$L = 10,000$ years

For most current estimates, we get a result that is in the tens or hundreds—but that's far fewer than earlier estimates, which were much higher: numbers in the 100,000s were often quoted. Today, we think it is more likely that tens or hundreds of intelligent races in this galaxy should be willing and able to communicate with us. Since the estimate for L is 10,000 years and we ourselves have been able to listen to interstellar messages for less than 100 years, we should expect most of these races to be far more advanced than we are.

That's good news for the "UFOs as aliens" enthusiasts, but there is a problem. The problem is quite simply that so far, we have detected no intelligent signals at all. That isn't entirely surprising, and it highlights a major problem. Since 1960, more than a hundred SETI searches have been able to look at only a tiny fraction of our galaxy; it won't be until the mid-2020s that (funding permitting) we will have a radio telescope capable of detecting signals from most of our galaxy. In the meantime, the Allen Telescope Array at Hat Creek in California will survey the galactic center for intelligent signals.

Finding intelligent life will be an enormous task.

Figure 14.1* shows a cluster of galaxies—any of which could well support intelligent life—imaged by the HST deep field telescope. This shows just a minute section of the sky, but it's probably typical of what the universe looks like in *any* direction. It seems certain that there is intelligent life somewhere out there, and civilizations far more advanced than ours—but the distances may be forever too great for us to communicate with (never mind visit) them, or for them with us.

Figure 14.1. Hubble Space Telescope image.

* Courtesy of NASA and NSSDC.

14.2 The Fermi Paradox

In 1960, Enrico Fermi, one of the twentieth century's greatest physicists, posed a lunchtime question to his colleagues when he asked, "Where *is* everybody?"

Fermi was wondering why it is that we haven't yet seen any evidence of alien civilizations. His calculations, worked out on the spot, suggested that Earth should have been contacted by alien civilizations many times in the past, and that we should see evidence of this.

The fact is—and this is the bad news for the "UFOs as aliens" enthusiasts—we don't.

It therefore seems likely that we have put some wrong numbers into the Drake Equation. Small changes make large differences to N. Perhaps the most problematic of all these values—the one that really can only be guessed at—is the lifetime of a civilization that would want to communicate with us. Mankind is the only example we have, and we have been in the interstellar communications business for less than 100 years. Atomic war, global warming, global cooling, asteroid impact—any of these things could cut short our term. That figure of 10,000 years could be wrong by a factor of 100. Or a value of 0.01 for f_i could be seriously wrong. Maybe a hundred to one against is way too high. Perhaps it takes a chance meteorite impact at *exactly* the right point in the evolution of life—in our case at the end of the cretaceous period—to switch evolving life away from armaments and defences towards intelligence.

Stephen Webb, in his excellent and erudite book *If the Universe is Teeming with Aliens, Where is Everybody?* (Copernicus, 2002), looks at the many and varied solutions to the Fermi Paradox. He divides the solutions into three groups—"They are here", "They exist but have not yet communicated", and "They do not exist". In his conclusion he begins by pointing out the only definite evidence that we have: to date, no alien intelligences have communicated with us or visited us. Moreover, there are no signs of their existence, despite SETI and other initiatives. We seem to be alone.

This might not be the case. There are all sorts of possible solutions to the Fermi paradox. Webb outlines no fewer than *50*. Maybe aliens are watching us, but we are so far below them in technology and intelligence, we can't tell. (Imagine yourself watching an ant colony. What chance have the ants of knowing you are looking at them, or of understanding what a human being is?) Maybe we are at a stage in our development where intelligent aliens are deliberately preventing our knowing what's going on out there (the quarantine hypothesis). Maybe aliens are beaming signals at us that we can't yet detect or understand. Maybe almost anything.

But at this point in time, the hard evidence for alien life in the universe is zero.

14.3 The Burden of Proof

When faced with zero evidence like this, the burden of proof is very much with the plaintiff. In legal terminology the "burden of proof" is the obligation in a particular context to defend a position against a different prima facie (at first sight) position. Here, the prima facie position is that there are no aliens because no aliens have ever been contacted or even detected by scientists.

If you see a UFO and want people to believe that it is connected in any way at all with alien life, you've got a great deal of work to do, beginning with proving "beyond reasonable doubt" that aliens are out there and in a position to communicate with us or observe us. The quote from Thomas Jefferson at the beginning of this chapter relates to this burden of proof: although he was wrong about meteorites, he was completely right about the burden of proof.

Observing a bright light in the sky that you can't identify and thinking that it might be a flying saucer isn't enough.

14.4 The Will to Believe

The expression "The Will to Believe" was coined by William James in his 1869 address to the Philosophical Clubs of Yale and Brown Universities. Something is deeply embedded in the human psyche that makes us able to believe, deeply and wholeheartedly, things for which there is no evidence at all. In the absence of evidence we look to gurus, leaders, or authority figures, and simply model our beliefs on theirs. We just accept what we've been told, read, or seen on television.

Why? There are a number of possibilities. Among the most compelling is that way back in mankind's distant past there was an evolutionary advantage in breaking large communities into smaller groups. If there was no real reason for a big community to separate into two smaller ones after a time, some invented reason became a factor that was important enough to disrupt the group. This could work *only* if the people on both sides of the divide believed it (the invented reason) unequivocally. Millennia later, we humans still have a predisposition to divide and even fight to the death over social matters that should seem to any rational mind to be unimportant or even silly. I rather favor this explanation, because it accounts for the lunacy of people fighting and dying in conflicts that have their basis in racial differences, religion, or nationality.

We have an inbuilt *will to believe*. This could be why there are people who *want* to believe that UFOs are evidence of alien visits to Earth, and hold on to their belief with a passion that can exceed the faith of the most fervent religious believers.

The will to believe clouds the picture when we are trying to determine the nature of a UFO. It makes it difficult to be sure of anyone's evidence about UFOs, because the witnesses may have evolved to be inherently unreliable. "UFOs as aliens" believers may be no more crazy than the vast number of people who believe in the revealed religions: there is in general no more and no less evidence for their faith. I guess it's just that they are in a minority.

Perversely, the "will to believe" can also include the will to believe that there are *no* aliens, and that we are alone in the universe. This is a perfectly arguable conclusion, but it is one that a rational person should reach only after examining all the evidence available, rather than leaping to it as an article of faith.

For my own part, I would love to see a UFO that I believe is an alien spacecraft. I would love to see a ghost. I would love to discover convincing evidence of an afterlife. So far I haven't, but I'll content myelf with having had the privilege of witnessing live television pictures of the first men walking on the Moon.

14.5 Abductees

There are many people—up to four million in the United States alone—who not only believe that aliens have contacted us, but that they personally have been abducted by aliens and later returned to Earth.

The most unexpected feature of these people as a group is that they don't show any signs of mental illness (no more than in the general population). They just sincerely believe that they were once abducted by aliens.

Psychology is a confusing (and from at least one astronomer's point of view, confused!) science, where very little can be said for sure, and where different ideas and explanations proposed by leading academics are very hard to prove one way or another or even to test. But progress has been made, progress that points to "alien abductees" having false memories, no surprise to most of us, given the lack of evidence for aliens at all.

A possible reason is a minor—and usually transient—problem the brain sometimes has when waking up. It is normal for the body to be effectively paralyzed—"sleep paralysis"—while we are asleep. Sleepwalking can be the result of the failure of this mechanism. But is possible for waking and the end of paralysis to get out of step by a minute or two. The sleeper then wakes up to find his body paralyzed. This state can be accompanied by "hypnopompic hallucinations"—a fairly common occurrence—that take (among others) the form of lights, a feeling of being watched, voices, and strange physical sensations. Upon waking up properly, the brain, which always seeks an interpretation of sensory input, can jump to the "alien abduction" conclusion as the most likely cause of the sensations. Voices, paralysis—which feels similar to being totally restrained on the bed—and strange lights make at least a kind of sense. The waking brain, and of course the person whose brain it is, has no way of knowing that the memories themselves were hallucinatory.

There are of course other hypotheses that psychologists have put forward to explain the phenomenon of alien abduction experiences, but here is at least one reasonable possibility. Given that the burden of proof of actual alien abduction is very substantial, the prima facie position must be to assume that some quirk of the human brain is responsible.

14.6 UFO Observations

At this point I want to look once again at—or maybe for—UFOs. There are a number of possibilities for ways in which a UFO might be observed. Broadly these are as follows:

1. Visual observation in which several people independently see an unidentified object in the daytime sky, and two or more of them photograph it.

2. Visual observation in which several people independently see an unidentified light in the night sky, and two or more of them photograph it.
3. Visual observation in which you see an unidentified object in the daytime sky and photograph it.
4. Visual observation in which you see an unidentified light in the night sky and photograph it.
5. Visual observation in which you see an unidentified object in the daytime sky.
6. Visual observation in which you see an unidentified light in the night sky.
7. UFOs that are photographed, but which nobody had seen until they looked at the photograph.
8. Hoaxes.

It may seem pedantic to list them all out like this, but it helps to sort them. The list is set out in order of credibility for any given sighting. It's more likely that an object, flying or otherwise, will be correctly identified in broad daylight than it will at night. The more the people who see it, especially if they are in different locations, the more likelihood there will be of one or more of them recognizing it as a helicopter, planet, or whatever so that it will cease to be unidentified.

If the observations—especially daytime ones—are supported by photographs, then there is the opportunity to analyze the sighting afterwards. So, on to the photographs themselves.

14.7 UFO Photographs

Modern digital cameras are capable of producing amazing results, compared with 35-mm film cameras. I still have a beautifully built and once-valuable Nikon F2 camera and some of the best lenses and accessories made for it, but I haven't used it for years. I also own three digital cameras, including a Fujicolor Finepix S7000, which has entirely supplanted my old F2.

"Darkroom" work on my F2 pictures was incredibly time-consuming, expensive, messy, and (in my case at least) prone to rather variable results. Processing digital photos on a PC is a joy in comparison, and all sorts of things are possible. The main problems with digital UFO photos today is that they are easy to fake, and even easier to mistake.

14.7.1 UFOs that Appear Unexpectedly on Photographs

Photographing a UFO by accident is easy to do. The problem is that you can't be sure what it is you've photographed.

I carried out some experiments for this chapter, and decided to photograph a helicopter at considerable distance to see how visible it would be. I used the Finepix S7000 without zooming (equivalent to a 50-mm standard lens on a 35-mm film camera), and the result is shown in Figure 14.2.

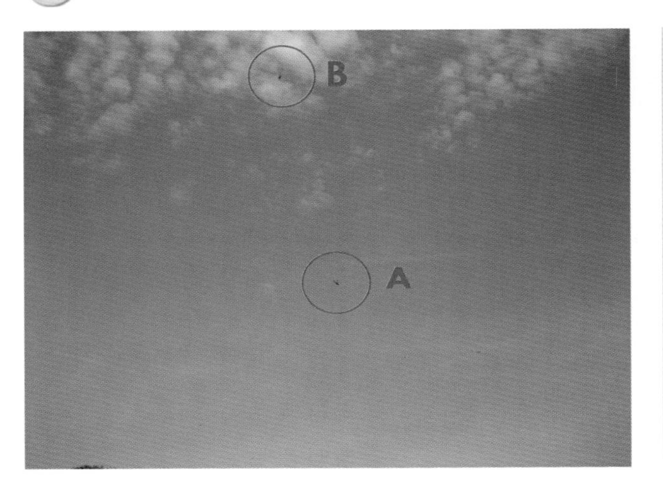

Figure 14.2. (Courtesy of John Watson).

It is the full frame, unprocessed (except for what the camera does automatically). The helicopter is circled and marked "A." When substantially enlarged, the helicopter is easily visible for what it is—once again, this is without any image processing except in the camera (Figure 14.3).

Figure 14.3. (Courtesy of John Watson).

But there are two objects in the photograph, the second one—marked "B"—was something I just didn't see when I took the picture. It is shown in Figure 14.4, at about the same magnification as the helicopter.

What can it be? Could I have inadvertently imaged an alien visitor? Sadly, no. It's a pigeon. Clearly it was much closer to me than the helicopter, which accounts for the motion-blurring. The wings look bigger than they are, are less dark than the bird's body because they were flapping and so part of the sky behind was captured during the exposure, even though it was a short exposure of 1/600th second—enough to "stop" the helicopter's rotor blades.

The warning is obvious: an image is dramatically affected by the way the camera works and may not easily be identifiable. But that doesn't make it an "alien" UFO.

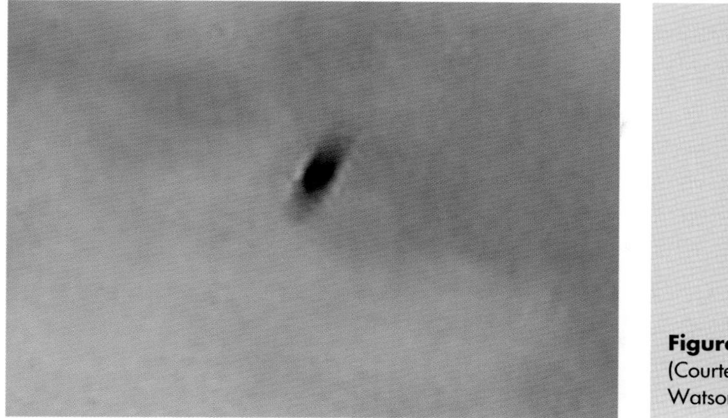

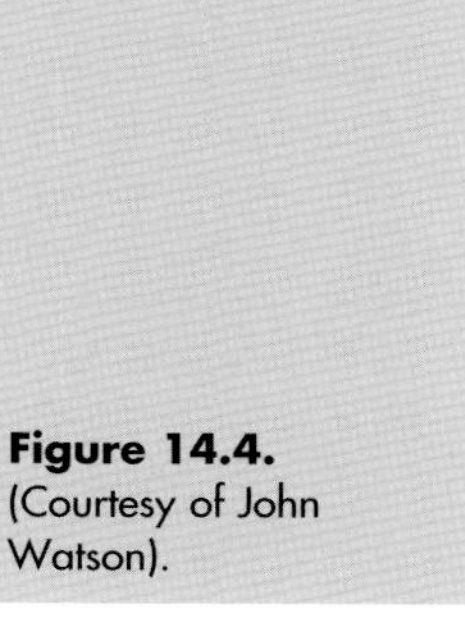

Figure 14.4. (Courtesy of John Watson).

Not only can the camera make images look unusual, there are actually some unusual things up there! Look at Figure 14.5.

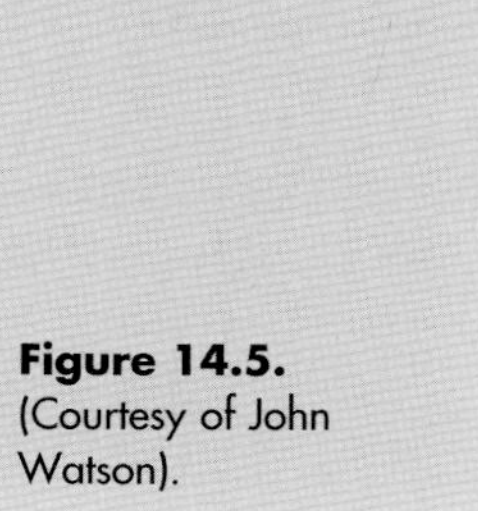

Figure 14.5. (Courtesy of John Watson).

The first impression is that Darth Vader's TIE-fighter has been photographed-flying over Hampshire, but once again this wasn't the case. It's an aircraft of course—a Bell Boeing Osprey V22. The Osprey has two enormous propellers on conventional wings. The engines, which are on the very tips of the wings, can be tilted in flight to allow the Osprey to land and take off vertically. A closer view is shown in Figure 14.6.

Through sheer chance, the camera shutter "stopped" the three-bladed propellers in a position that makes the aircraft look like a TIE-fighter when seen from the rear (or front). This photograph was taken with a Canon Powershot SD450 (it's called an "Ixus 55" in the United Kingdom, for some reason) at 1/1100th second exposure.

Figure 14.6. (Courtesy of John Wastson).

14.7.2 Nighttime UFO Photographs

This category of UFO is more problematic, because basically you are photographing lights, usually at some distance. The results depend mostly on the camera, photographic technique, and some foreknowledge of what you're photographing. The shot in Figure 14.7 shows a mysterious "flying tick" over my garden.

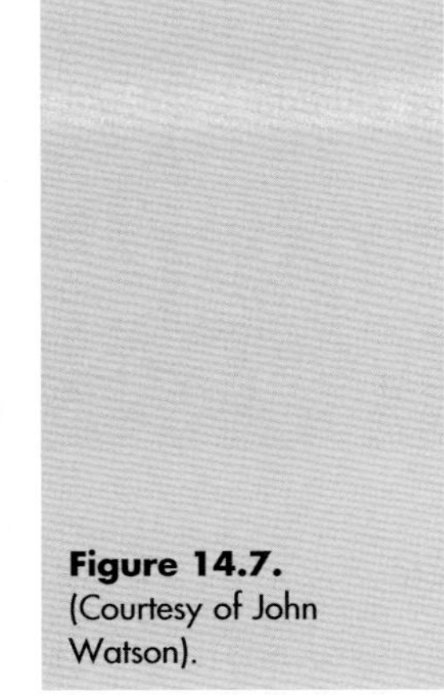

Figure 14.7. (Courtesy of John Watson).

It's actually Jupiter, down low in the south. It was photographed with the Finepix S7000 at full zoom, ISO 400, and with a 1.5× teleconverter, the equivalent of about a 750-mm lens on a 35-mm film camera. I handheld the camera for the 1/5th second exposure, and the result was serious camera shake.

I tried again when it was a little darker. This time it was the same camera, lens, and setup but with an exposure of 3 s. The "knitting" which resulted is probably familiar to most astronomers (Figure 14.8).

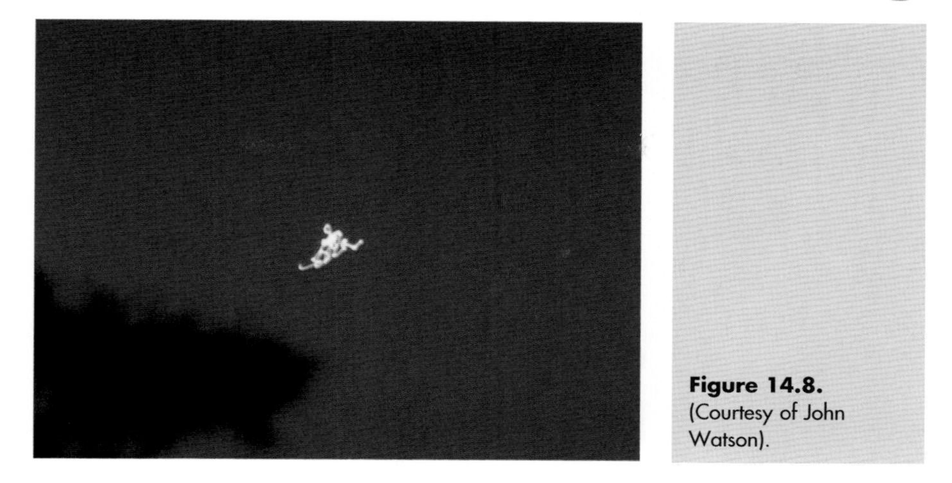

Figure 14.8. (Courtesy of John Watson).

Finally, we have Jupiter in (Figure 14.9)—the same lens and setup again, but with the camera on a tripod and the self-timer to allow vibrations to die down.

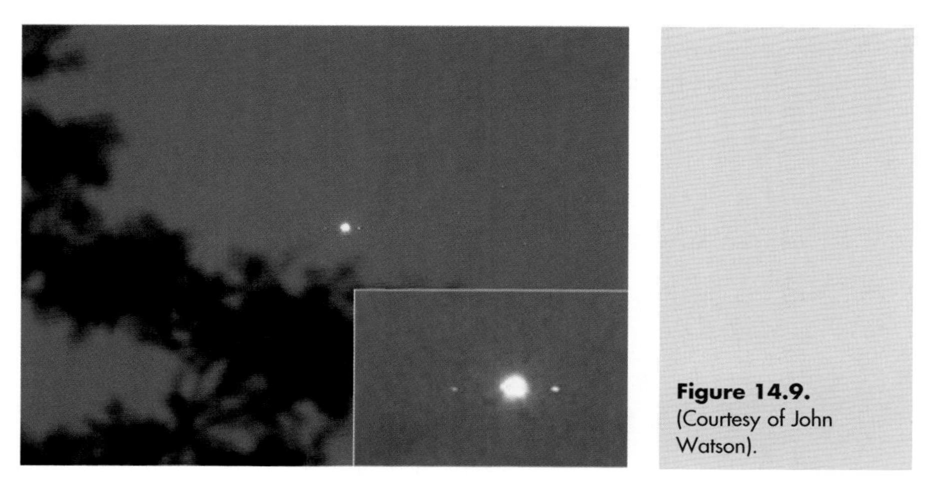

Figure 14.9. (Courtesy of John Watson).

I think it's a tribute to the camera that at least two of the Galilean satellites are visible in the picture—Io and Ganymede—and possibly even Europa just at the planet's western (left in the photo) limb. (The inset picture shows a magnified view.)

14.7.3 Optical Effects

There are many different optical effects that can result in spurious images or distort real ones. Aim your camera at a bright light and you're almost certain to see something. Dirty filters or shooting through a dirty window can cause odd spots and dots. Any one of these should be pretty obvious to an amateur astronomer, so I don't propose going into any detail.

(A)

(B)

Figure 14.10. *Glory or UFO at LA 1975* These bright lights in the sky are glories caused by sunlight reflected off the aircraft windows.

14.7.4 Videos

More or less the same caveats apply to videos of UFOs. The way the frame is scanned can result in some special kinds of image distortion—the primary explanation for the "alien rods" videos of 1999. Flying insects appear elongated because of the camera's image scanning, making them look rod-like. It got a TV program, but that doesn't make it anything other than an artifact of the camera.

14.8 Sightings of "Alien" UFOs at Night

People sometimes report bright lights that follow their car or airplane. These tend to be explained away as "probably Venus" for no reason other than the fact that Venus is very bright and might well have been low in the sky and in the right direction. It's an explanation that may sound facile, but it is often right. What people tend not to appreciate is the degree to which the human brain, striving to find patterns in everything, processes the optical information from our eyes. Think about driving fast at night. Scenery is rushing past you at great speed, but with very little visual input you drive with complete confidence. Based on no more than a glimpse of a fence here, maybe a half-seen road marking or flickers of light in your headlight beam, or distant tail-lights, your brain confidently models the road ahead with stunning accuracy. Humans are incredibly good at mentally reconstructing the landscape in front of them at night, from the tiniest scraps of evidence. No computer could even begin to match us in it.

But we are not perfect. Sometimes our brain makes a mistake, particularly where an unusual visual input is involved. A bright light—like the planet Venus—can get incorporated into the moving model that the brain calculates to be a true representation of the road ahead. It is possible to be utterly confident that "the light was following me, staying just ahead" because you *have* to be confident in your brain's model of the scene in front of you. If you weren't, night driving would involve inching along at a snail's pace, just in case.

You'll even recall, as clearly as anything else, that the light was moving; that's because the parts of your brain that constitute your intellect have no way of knowing that your visual processing system—which works automatically all the time—made a rare mistake.

14.9 Sightings of "Alien" UFOs in the Daytime

There are many recorded apparent visual sightings of "alien" UFOs in daylight. One of the most persuasive, because it was so thoroughly investigated, was the McMinnville (Oregon) sighting of 11 May 1950. Two witnesses reported seeing a huge saucer-shaped craft flying past their backyard, south of the Salmon River Highway.

The Condon Report[†]. concluded that "This is one of the few UFO reports in which all the factors investigated, geometric, psychological and physical, appear to be consistent with the assertion that an extraordinary flying object, silvery, metallic, disk-shaped, tens of meters in diameter, and evidently artificial, flew within sight of two witnesses."

From the vantage point of 40 years later, many people are now not so sure. Look on the Internet to see the various arguments pro and con. The report's conclusions hinged on a detailed photometric analysis of the photographs, and on the hitherto impeccable characters of the witnesses. Robert Sheaffer's paper of 1969 (updated in 1999 and available on the Internet) came to a conclusion that the Condon investigation's photometry was at best questionable and probably wrong, which leaves just the two witnesses. . . .

So would the McMinnville sighting pass the test of Burden of Proof? Is it an alien vehicle "beyond reasonable doubt"?

Of course not.

Which brings us neatly round to hoaxes.

If someone is trying to fool you, then by definition they will try to make it seem like a "real" alien UFO. Hoaxes are most often photographs. Some are obvious, some less so. Some involve two or more people confirming the story. None that I know of could even get close to passing the test of "beyond reasonable doubt".

14.10 Amateur Astronomers and UFOs

As an astronomer, what should you do when someone excitedly points out a UFO in the night sky?

I can only say what *I* do in such circumstances. Get the person to *look* at the object. No, *really* look at it. Suggest all the things an unidentified light in the sky might be: an artificial satellite or booster debris, a high-flying aircraft, a meteor, luminous cloud, ice crystals, plasmas, a balloon, all manner of possibilities. If they still think it's more likely to be an alien mother-ship, talk about the burden of proof. That should be enough.

If it isn't, get them to read this book!

[†] *Final Report of the Scientific Study of Unidentified Flying Objects*, conducted by the University of Colorado under contract to the USAF. The Scientific Director was Dr. Edward U. Condon. © 1968 The Board of Regents of the University of Colorado. Published by the New York Times Book Company, 1969.

CHAPTER FIFTEEN

Oddities

There are always some lights in the sky that don't fit into any logical scheme. No doubt there are plenty of others, some yet to be properly documented. These few are now better understood and don't puzzle as they used to in days of yore.

Definitions of phosphorescence, fluorescence, and luminescence in the sky are given in Chapter 10. Here a few others of natural origin are discussed.

15.1 Fireflies and Gloworms

In this category of insect lights, we can't really consider gloworms as true lights in the sky as the female is earthbound. The male seeks out these fascinating lights in the tail. All such earthbound lights are regarded as a special case when seen against or on a grassy bank.

The famous gloworm caves in New Zealand when seen from below at ground level do come into a category of lights in the sky in the huge caverns. Here the insect is a fly, not a beetle, but that's how it's known. It clings to the cavern roof and spins a gossamer thread, very similar in nature to a spider's. At rest, or when well fed, the tail of the insect is extinguished, or dimly lit.

Something magical happens when the insect is hungry. The brightness increases considerably but that is not all the magic—the light now shines at the end of the trailing thread as a really intense blue spot, a considerable distance below the cavern roof. This will attract its prey, small flying insects, which are wound in when ensnared. These threads are a natural fiber optic, predating man's efforts by millions of years.

It is only when one sees this myriad of lights in a vast cavern that the sheer wonder of these true lights in the sky is fully appreciated. The impression of a starlit sky is never to be forgotten, definitely something in "What to Observe."

Worldwide, there are a host of fish that glow, and many are seen at or close to the surface. Cuttlefish and some squid can glow, and that's one way of attracting them to lamps held over a boat's gunnel.

More intriguingly, the bulk of light and glow seen at ocean surface levels comes from microscopic plankton, not just a single species but umpteen varieties, some occurring in synchronicity or in myriad species. Ocean travelers over the ages have used that light as a navigational aid to travel in what would otherwise be total darkness on a moonless or cloudy night.

Many environmental experiments are now conducted from satellites that specifically monitor this elusive natural light from the plankton and other living ocean flora. The same craft also monitor the unwelcome glow from pollution.

This form of light becomes a glow in the sky when it's reflected off cloud, or any mist or particulate matter. Navigators knew they were in the wide ocean if the sky glowed all around. Any dark patch was worth investigating as a possible landfall.

The true night skylights are fireflies. There are a bewildering number, most in huge numbers in the tropics.

The mechanism by which they glow, which is a neat piece of chemistry involving little or no heat, is fully understood today. The basic chemicals can be bought and assembled on a DIY basis.

All the safelights and toys seen at carnivals are based on that neat piece of natural chemistry. Next time you buy a glowstick, think of the origins in some extremely efficient insect and animal chemistry.

All that we do is add a few extra colors, and virtually, the whole spectrum is now sold.

15.2 Will-O-The-Wisp

Treated with great respect or even fear in ancient times, this transient glow waited a long time for an explanation. The basic fear arose because it was normally only seen in deadly bogs or churchyards, anywhere, in fact, associated with death and destruction.

Some phosphorescence associated with marsh gas, the ubiquitous methane, is the source of the light, which can glow quite strongly on occasions.

Some suggestions indicate that the stuff can ignite. Much more likely is an association with true phosphorescence from the phosphorus in rotting corpses and similar decay.

Whatever the true source of the ghostly light was in those days, some fungi do emit light, though not very strongly in most cases. Some of these also have a pungent odor and that can also lend itself to an association with rotting. When seen on branches and similar outcrops, they are valid inclusions as lights in the sky.

15.3 Radioactivity

This is one glow in the dark you don't want to encounter at ground level. Radioactive decay forms part of the nightglow, Chapter 10. Some of the early experiments studied that natural phosphorescence in minerals and chemicals to sort out the basic physics. Some of the radium collected by the Curies did glow, and Czerenkov radiation has been mentioned many times already.

At that time, the only way to detect radiation was with a screen coated in some phosphor. Alpha radiation was seen as discrete spots or flashes.

We don't need to define radioactivity here, the point being made and to bear in mind is very simple: That phenomenon led directly to our fluorescent lighting, thence to all our modern energy saving lamps. All TV screens worked that way, too, until LCDs arrived. Plasma screens are related. The unwanted side effect from all that modern technology is light pollution, discussed in Chapter 3.

15.4 Ultraviolet

A familiar party trick is to turn on an ultraviolet light and show up a wide variety of bright colors. Whiter-than-white additives in washing powders have largely gone out of favor, but the quinine in bitter lemon and similar drinks will glow extremely brightly.

The importance of this effect cannot be overemphasized since it is the basis of many of the upper atmospheric glows and whenever someone turns on a bright lamp. This effect can be mimicked—always be wary of strange glows in the sky when anywhere close to a town or city. It could well be an advertising gimmick from some supermarket.

Figure 15.1. *Grot on window* Photographing through dirty windows can create some "odd" lights in the sky. Pure fun.

15.5 True Oddities

When all is said and done, there will always be a residue of interesting sights in the sky that we can admire or ponder over. Many of them will be a variant of the intriguing lights and glows in the sky I've tried to bring to your attention.

Figures 15.1 and 15.2 are merely a begining of some ideas where you can create your own through some lateral thinking.
If I've inspired you to go out and have a look, so much the better.

HAPPY HUNTING

Figure 15.2. *Blue moon to order* Poorly adjusted "white balance" might explain many other "odd" lights in the sky. Reflections from double glazing also illustrate how moon and sunlight exposure can be brought under control.

Artistry Combined with Science

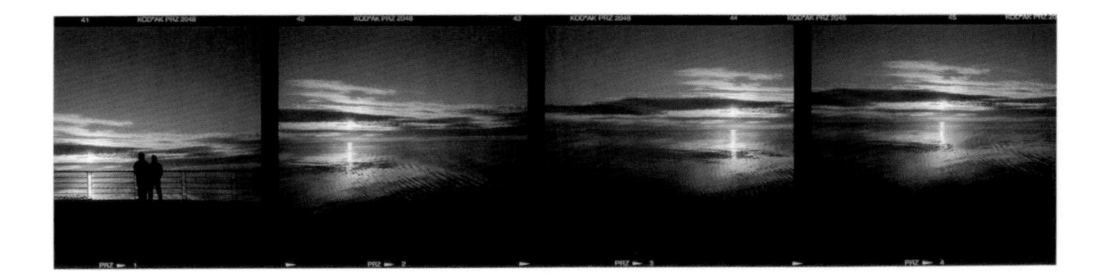

X.1. Antarctica midnight sun sequence.

X.2. Arc and sundial.

X.3. Chile eclipse, 1994.

X.9. Lone rhino reflection.
Note: the purple reflection due to volcanic dust.

X.10. Lunar corona over St. Pauls, London, 1970.

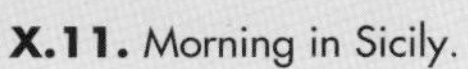

X.11. Morning in Sicily.

X.13. Unusual clouds, Scott Base, 1998.

X.14. Unusual street lighting, Santiago. Note: the Moon does not have a halo.

X.15. Unusual sunset.

X.17. Very clear skies and cactus Galapagos.

Index

Other Titles in this Series

(Continued from page ii)

The New Amateur Astronomer
Martin Mobberley

Lunar and Planetary Webcam User's Guide
Martin Mobberley

Choosing and Using a Schmidt-Cassegrain Telescope
A Guide to Commercial SCT's and Maksutovs
Rod Mollise

The Urban Astronomer's Guide
A Walking Tour of the Cosmos for City Sky Watchers
Rod Mollise

Astronomy with a Home Computer
Neale Monks

More Small Astronomical Observatories
Sir Patrick Moore (Ed.)

The Observer's Year
366 Nights in the Universe
Sir Patrick Moore (Ed.)

A Buyer's and User's Guide to Astronomical Telescopes and Binoculars
Jim Mullaney

Care of Astronomical Telescopes and Accessories
A Manual for the Astronomical Observer and Amateur Telescope Maker
M. Barlow Pepin

The Deep-Sky Observer's Year
A Guide to Observing Deep-Sky Objects Throughout the Year
Grant Privett and Paul Parsons

Software and Data for Practical Astronomers
The Best of the Internet
David Ratledge

Digital Astrophotography: The State of the Art
David Ratledge (Ed.)

Spectroscopy: The Key to the Stars
Keith Robinson

CCD Astrophotography:High-Quality Imaging from the Suburbs
Adam Stuart

The NexStar User's Guide
Michael Swanson

Astronomy with Small Telescopes
Up to 5-inch, 125 mm
Stephen F. Tonkin (Ed.)

AstroFAQs
Questions Amateur Astronomers Frequently Ask
Stephen F. Tonkin

Binocular Astronomy
Stephen F. Tonkin

Practical Amateur Spectroscopy
Stephen F. Tonkin (Ed.)

Amateur Telescope Making
Stephen F. Tonkin (Ed.)

Using the Meade ETX
100 Objects You Can Really See with the Mighty ETX
Mike Weasner

Observing the Moon
Peter T. Wlasuk

Printed in Thailand